Biodiversity Under Threat

ISSUES IN ENVIRONMENTAL SCIENCE AND TECHNOLOGY

EDITORS:

R.E. Hester, University of York, UK
R.M. Harrison, University of Birmingham, UK

EDITORIAL ADVISORY BOARD:

Sir Geoffrey Allen, Executive Advisor to Kobe Steel Ltd, UK, **A.K. Barbour**, Specialist in Environmental Science and Regulation, UK, **P. Crutzen**, Max-Planck-Institut für Chemie, Germany, **S.J. de Mora**, Aromed Environmental, Kingston, Canada, **G. Eduljee**, SITA, UK, **J.E. Harries**, Imperial College of Science, Technology and Medicine, UK, **S. Holgate**, University of Southampton, UK, **P.K. Hopke**, Clarkson University, USA, **Sir John Houghton**, Meteorological Office, UK, **P. Leinster**, Environment Agency, UK, **J. Lester**, Imperial College of Science, Technology and Medicine, UK, **P.S. Liss**, School of Environmental Sciences, University of East Anglia, UK, **D. Mackay**, Trent University, Canada, **A. Proctor**, Food Science Department, University of Arkansas, USA, **D. Taylor**, AstraZeneca plc, UK, **J. Vincent**, School of Public Health, University of Michigan, USA.

TITLES IN THE SERIES:

How to obtain future titles on publication

A subscription is available for this series. This will bring delivery of each new volume immediately on publication and also provide you with online access to each title via the Internet. For further information visit http://www.rsc.org/Publishing/Books/issues or write to the address below.

For further information please contact:
Sales and Customer Care, Royal Society of Chemistry, Thomas Graham House, Science Park, Milton Road, Cambridge, CB4 0WF, UK
Telephone: +44 (0)1223 432360, Fax: +44 (0)1223 426017, Email: sales@rsc.org

ISSUES IN ENVIRONMENTAL SCIENCE AND TECHNOLOGY

EDITORS: R.E. HESTER AND R.M. HARRISON

25
Biodiversity Under Threat

RSCPublishing

ISBN: 978-0-85404-251-7

ISSN: 1350-7583

A catalogue record for this book is available from the British Library

Published by The Royal Society of Chemistry,
Thomas Graham House, Science Park, Milton Road,
Cambridge CB4 0WF, UK

Registered Charity Number 207890

For further information see our web site at www.rsc.org

Preface

Biodiversity has become quite a buzzword of late. Hardly a day goes by without alarmed reference to it in the popular media. Much of the current comment is linked to climate change, widely believed to be caused by human activities. Other anthropogenic causes of biodiversity loss, such as deforestation, over-fishing, intensification of agriculture, pollution and the spread of invasive species also have received considerable attention. This book brings together an international group of experts on the subject, each with a distinctive focus, giving an overview of its many different aspects and combining academic rigour with a concern to make the topic intelligible to the non-specialist reader.

Biodiversity plays an important role in the sustainability of ecosystems and provides both goods and services that are essential for human survival. However, despite the increased awareness of its benefits, biodiversity is undoubtedly under threat from the many pressures imposed by human-induced changes. The ten chapters of this book provide a broad view of the many threats to global biodiversity and of the policy responses required to combat them. Thus policy is a theme common to several of the chapters, but for the most part this is dealt with in the specific context of the particular topic under discussion, *e.g.* invasive species, threatened habitats, land use change, *etc.*

The book begins with a chapter by Nigel Boatman and his colleagues from the UK government's Central Science Laboratory. This is concerned with the impacts of agricultural change on farmland biodiversity in the UK. Next Jessica Hellmann of the University of Notre Dame and Nathan Sanders of the University of Tennessee, USA, write on the extent and future of global insect biodiversity. Chapter 3 then considers biological invasions in Europe, with an analysis of the relevant pressures, states, impacts and responses by Philip Hulme, now at the National Center for Advanced Bio-Protection Technologies in Lincoln University, NZ. In Chapter 4 Paul Tyler of the UK National Oceanography Centre in Southampton writes on biodiversity in the deep sea, addressing the question "if we do not understand the biodiversity, how can we assess the threat?" The fifth chapter, by Alison Hester and Rob Brooker of the Macaulay Institute in Aberdeen, Scotland, addresses threatened habitats, with a focus on marginal vegetation in upland areas. It is from this chapter that we draw the illustration used on our front cover.

The second half of the book begins with a chapter on trends in biodiversity in Europe and the impact of land use change, written by Allan Watt of the CEH, Banchory, Scotland, together with co-authors from Denmark, France, Ireland, England, Spain, Germany, Sweden, Hungary, Finland, The Netherlands and Portugal. Then in Chapter 7 Jon Lovett and colleagues from the Institute for Tropical Ecosystem Dynamics in York, England, provide an account of tropical forest biodiversity. Chapter 8 deals with both constraints and successes in the implementation of international biodiversity initiatives and is written by Eeva Furman of the Finnish Environment Institute in Helsinki, together with co-authors who include colleagues from both The Netherlands and Romania. In Chapter 9, written by another international team headed by Michael Bredemeier of the Forest Ecosystems Research Centre at the University of Gottingen, Germany, and including co-authors from Wales, Hungary, Austria and Italy, the subject of biodiversity assessment and change and the challenge of appropriate methods are addressed. Finally, the book closes with a chapter by Stefan Klotz of the UFZ-Centre for Environmental Research in Lepzig-Halle, Germany, on the role of natural and anthropogenic drivers and pressures on biodiversity.

The subject of biodiversity is, of course, huge in scope and significance. Even in 10 chapters, involving some 47 authors, we cannot claim a fully comprehensive treatment. However, we believe the current understanding of the threats to biodiversity is particularly well described here, with a wide range of illustrative examples. Future directions for increasing this understanding and developing appropriate policy initiatives to combat the worst of the threats are suggested and discussed. Thus the book will be of value to policymakers as well as to ecologists and environmental scientists and to all students of the environment.

Ronald E Hester
Roy M Harrison

Contents

Impacts of Agricultural Change on Farmland Biodiversity in the UK
Nigel D. Boatman, Hazel R. Parry, Julie D. Bishop and Andrew G.S. Cuthbertson

The Extent and Future of Global Insect Diversity
Jessica J. Hellmann and Nathan J. Sanders

Biological Invasions in Europe: Drivers, Pressures, States, Impacts and Responses
Philip E. Hulme

The Deep Sea: If We Do Not Understand the Biodiversity, Can We Assess the Threat?
Paul Tyler

Threatened Habitats: Marginal Vegetation in Upland Areas
Alison Hester and Rob Brooker

Trends in Biodiversity in Europe and the Impact of Land-use Change
A.D. Watt, R.H.W Bradshaw, J. Young, D. Alard, T. Bolger,
D. Chamberlain, F. Fernández-González, R. Fuller, P. Gurrea, K. Henle,
R. Johnson, Z. Korsós, P. Lavelle, J. Niemelä, P. Nowicki, M. Rebane,
C. Scheidegger, J.P. Sousa, C. Van Swaay and A. Vanbergen

Tropical Moist Forests
Jon C. Lovett, Rob Marchant, Andrew R. Marshall and Janet Barber

The Implementation of International Biodiversity Initiatives: Constraints and Successes

Eeva Furman, Riku Varjopuro, Rob Van Apeldoorn and Mihai Adamescu

Biodiversity Assessment and Change – the Challenge of Appropriate Methods

Michael Bredemeier, Peter Dennis, Norbert Sauberer, Bruno Petriccione, Katalin Török, Cristiana Cocciufa, Giuseppe Morabito and Alessandra Pugnetti

**Drivers and Pressures on Biodiversity in Analytical
Frameworks**
Stefan Klotz

Editors

Ronald E. Hester, BSc, DSc(London), PhD(Cornell), FRSC, CChem

Ronald E. Hester is now Emeritus Professor of Chemistry in the University of York. He was for short periods a research fellow in Cambridge and an assistant professor at Cornell before being appointed to a lectureship in chemistry in York in 1965. He was a full professor in York from 1983 to 2001. His more than 300 publications are mainly in the area of vibrational spectroscopy, latterly focusing on time-resolved studies of photoreaction inter-mediates and on biomolecular systems in solution. He is active in environmental chemistry and is a founder member and former chairman of the Environment Group of the Royal Society of Chemistry and editor of 'Industry and the Environment in Perspective' (RSC, 1983) and 'Understanding Our Environment' (RSC, 1986). As a member of the Council of the UK Science and Engineering Research Council and several of its sub-committees, panels and boards, he has been heavily involved in national science policy and administration. He was, from 1991 to 1993, a member of the UK Department of the Environment Advisory Committee on Hazardous Substances and from 1995 to 2000 was a member of the Publications and Information Board of the Royal Society of Chemistry.

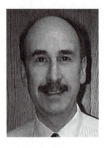

Roy M. Harrison, BSc, PhD, DSc(Birmingham), FRSC, CChem, FRMetS, Hon MFPH, Hon FFOM

Roy M. Harrison is Queen Elizabeth II Birmingham Centenary Professor of Environmental Health in the University of Birmingham. He was previously Lecturer in Environmental Sciences at the University of Lancaster and Reader and Director of the Institute of Aerosol Science at the University of Essex. His more than 300 publications are mainly in the field of environmental chemistry, although his current work includes studies of human health impacts of atmospheric pollutants as well as research into the chemistry of pollution phenomena. He is a past Chairman of the Environment Group of the Royal Society of Chemistry for whom he has edited 'Pollution: Causes, Effects and Control' (RSC, 1983; Fourth Edition, 2001) and 'Understanding our Environment: An Introduction to Environmental Chemistry and Pollution' (RSC, Third Edition, 1999). He has a close interest in scientific and policy aspects of air pollution, having been Chairman of the Department of Environment Quality of Urban Air Review Group and the DETR Atmospheric Particles Expert Group as well as a member of the Department of Health Committee on the Medical Effects of Air Pollutants. He is currently a member of the DEFRA Air Quality Expert Group, the DEFRA Advisory Committee on Hazardous Substances and the DEFRA Expert Panel on Air Quality Standards.

Contributors

Mihai Adamescu, University of Bucharest, Department of Systems Ecology, Spl. Independentei 91–95, 050095 Bucharest, Romania

D. Alard, Université Bordeaux 1, Bâtiment B8-RdC-Porte 01, Avenue des Facultés, 33405 Talence, France

Rob van Apeldoorn, Wageningen UR, Alterra, Landscape Centre, PO Box 47, 6700 AA Wageningen, The Netherlands

Janet Barber, Bovey Cottage, Caudle Green, Cheltenham, GL53 9PR, England, UK

Julie D. Bishop, Central Science Laboratory, Sand Hutton, York, YO41 1LZ, England, UK

Nigel D. Boatman, Central Science Laboratory, Sand Hutton, York, YO41 1LZ, England, UK

T. Bolger, Dept. of Zoology, University College Dublin, Belfield, Dublin 4, Ireland

R. H. W. Bradshaw, Department of Environmental History and Climate Change, 2400 Copenhagen NV. Denmark

Michael Bredemeier, University of Goettingen, Forest Ecosystems Res. Ctr., Buesgenweg 2, D-37077 Goettingen, Germany

Rob Brooker, Macaulay Institute, Craigiebuckler, Aberdeen, AB15 8QH, Scotland, UK

D. Chamberlain, British Trust for Ornithology, BTO, The Nunnery, Thetford, Norfolk, IP24 2PU, England, UK

Cristiana Cocciufa, Italian Forest Service, CONECOFOR Office, via Carducci 5 00187 Roma, Italy

Andrew G. S. Cuthbertson, Central Science Laboratory, Sand Hutton, York, YO41 1LZ, England, UK

Peter Dennis, Institute of Rural Sciences, University of Wales, Aberystwyth, SY23 3AL, Wales, UK

F. Fernández-González, Facultad de CC del Medio Ambiente. Avda. Carlos III, s/n. 45071 Toledo, Spain

R. Fuller, Dept. of Zoology, University College Dublin, Belfield, Dublin 4, Ireland

Eeva Furman, Finnish Environment Institute, Mechelininkatu 34 a, Helsinki 00930, Finland

P. Gurrea, Facultad de Biologýá Universidad Complutense, E-28040 Madrid, Spain

Jessica J. Hellmann, Department of Biological Sciences, University of Notre Dame, Notre Dame, IN 46556, USA

K. Henle, Department of Conservation Biology and Natural Resources, UFZ, Permoserstraße 15, 04318 Leipzig, Germany

Alison Hester, Macaulay Institute, Craigiebuckler, Aberdeen, AB15 8QH, Scotland, UK

Philip E. Hulme, National Center for Advanced Bio-Protection Technologies, PO Box 84, Lincoln University, Canterbury, New Zealand

R. Johnson, Agricultural Sciences, PO Box 7050, Vallvagen 3, Uppsala, Sweden

Stefan Klotz, Department of Community Ecology, Helmholtz-Center of Environmental Research – UFZ, Theodor-Lieser-Straße 4, D-06120 Halle/Saale, Germany

Z. Korsós, Hungarian Natural History Museum, Baross utca 13, Budapest H-1088, Hungary

P. Lavelle, IRD/Université de Paris VI, France

Jon C. Lovett, Centre for Ecology, Law and Policy, Environment Department, University of York, York, YO10 5DD, England, UK

Rob Marchant, York Institute for Tropical Ecosystem Dynamics, Environment Department, University of York, York, YO10 5DD, England, UK

Andrew R. Marshall, Centre for Ecology, Law and Policy, Environment Department, University of York, York, YO10 5DD, England, UK

Giuseppe Morabito, National Research Council, Institute for Ecosystem Research, Verbania, Italy

J. Niemelä, Department of Ecology & Systematics, University of Helsinki, 17 Arkadiankatue 7, 00014 Helsinki, Finland

P. Nowicki, ECNC, Fons vd Heydenstraat 57, NL-5534- AT Netersel, The Netherlands

Hazel R. Parry, Central Science Laboratory, Sand Hutton, York, YO41 1LZ, England, UK

Bruno Petriccione, Italian Forest Service, CONECOFOR Office, via Carducci 5 00187 Roma, Italy

Alessandra Pugnetti, National Research Council, Institute for Marine Science, Venice, Italy

M. Rebane, English Nature, Northminster House, Peterborough, PE1 1UA, England, UK

Nathan J. Sanders, Department of Ecology and Evolutionary Biology, University of Tennessee, Knoxville, TN 37996, USA

Norbert Sauberer, VINCA – Vienna Institute for Nature Conservation & Analyses, Giessergasse 6/7, A-1090 Vienna, Austria

C. Scheidegger, Swiss Federal Institute for Forest, Snow and Landscape Research, CH-8903 Birmensdorf, Switzerland

J. P. Sousa, Instituto do Ambiente e Vida, University of Coimbra, Portugal

C. van Swaay, Dutch Butterfly Conservation/De Vlinderstichting, Postbus 506 6700 AM Wageningen, The Netherlands

Katalin Török, Department of Plant Ecology, Institute of Ecology and Botany, Hungarian Academy of Sciences, Alkotmány u. 2-4, H-2163 Vácrátót, Hungary

Paul Tyler, National Oceanography Centre, University of Southampton, European Way, Southampton, SO14 3ZH, England, UK

A. Vanbergen, Centre for Ecology and Hydrology, Hill of Brathens, Banchory, AB31 4BW, Scotland, UK

Riku Varjopuro, Finnish Environment Institute, Mechelininkatu 34 a, Helsinki 00930, Finland

A. D. Watt, Centre for Ecology and Hydrology, Hill of Brathens, Banchory, AB31 4BW, Scotland, UK

J. Young, Centre for Ecology and Hydrology, Hill of Brathens, Banchory, AB31 4BW, Scotland, UK

Impacts of Agricultural Change on Farmland Biodiversity in the UK

NIGEL D. BOATMAN, HAZEL R. PARRY, JULIE D. BISHOP AND
ANDREW G.S. CUTHBERTSON

1 Introduction

Over the past 50 years, there has been a marked decline shown by many species
closely associated with lowland farmland in the UK, which is widely considered
to be a key issue in British nature conservation. Increased availability of survey
data has meant it is now possible to quantify changes in biodiversity for some
groups. Changes have occurred in many farming practices and these have
affected biodiversity in a variety of ways. Mixed agriculture in Britain has been
lost and farms have become specialised; traditional crop rotations have declined and pastoral and arable farming have become polarised. Farming has
intensified; for example, wheat yields in Scotland increased by 201% during the
period 1967–1999 due to more effective tillage, application of fertilisers and
pesticides and plant breeding. Field sizes have increased, reducing non-crop
habitat at field margins. Other changes include more autumn sowing of crops
and more efficient harvesting, more non-inversion tillage, drainage and reseeding of grassland, a switch from hay to silage, increased stocking rates and the
use of avermectin wormers.[1] This has all impacted upon wildlife species that
inhabit lowland farmland landscapes. Upland farming and livestock grazing
has also intensified during this time. High grazing pressure, especially by sheep,
has had a negative impact upon vegetation and wildlife in many upland regions
of Britain.[2]

Much of the focus on biodiversity conservation within agricultural landscapes has been on the conservation of rare or rapidly declining species, driven
in part by the UK Biodiversity Action Plan, the Government's response to the
1992 Convention on Biological Diversity, and also by the adoption of a
commitment as part of the Government's Public Service Agreement to "care

Issues in Environmental Science and Technology, No. 25
Biodiversity Under Threat
Edited by RE Hester and RM Harrison
© The Royal Society of Chemistry, 2007

for our natural heritage, make the countryside attractive and enjoyable for all and preserve biological diversity by reversing the long-term decline in the number of farmland birds by 2020, as measured annually against underlying trends; and bringing into favourable condition by 2010 95 per cent of all nationally important wildlife sites". However, other issues of importance in the context of agriculture include whether or not increased biodiversity or species richness enhances ecosystem functions such as primary productivity and nutrient retention or ecosystem services such as pollination and biological control.[3] Non-crop habitats on farmland are usually more species diverse than cropped fields or intensive grasslands, and the areas of non-crop habitat may even become islands of species richness if dispersal across suitable habitat is limited. However, over-zealous "tidying" of non-crop areas can reduce value for biodiversity and there may be a conflict here between the concept of an attractive (which for many equates with "tidy") countryside and the wish to see it populated by diverse fauna and flora.

A review of changes in biodiversity on arable farmland[4] concluded that around half of plant species, a third of insect species and four-fifths of bird species characteristic of farmland have declined. This chapter reviews the key changes that have occurred in agriculture during the second half of the twentieth century, set in the context of evolving policies on agricultural support and their impact on the characteristic fauna and flora of the farmed landscape. It concludes by considering the most recent reform of the EU Common Agricultural Policy and the implications of the shift away from production support and towards greater provision of incentives for environmentally sustainable management. A.D. Watt *et al.* present additional related and complementary material in Chapter 6 of this volume, with particular emphasis on the wider European aspects of land use change.

2   The Post-war Intensification of Agriculture

During and following the Second World War, agricultural policy focused on maximising food production. This resulted in an unprecedented level of change and intensification accompanied by subsidies ensuring prolonged price stability throughout the second half of the twentieth century. Although highly successful, this policy has been blamed for large declines in biodiversity within both the UK[4,5] and Europe.[6] Farming practices became polarised, where arable farming dominated the east of the UK and grassland/livestock farming the west, whilst mixed farming and the use of grass leys declined.

The key changes in both arable and grassland landscapes have been identified.[1] In arable landscapes, the key changes that took place in the last forty years include simpler rotations and block cropping; a switch from spring to autumn cropping; more efficient harvesting and sealed grain storage; and recently, more non-inversion tillage. On grasslands, key changes in the last forty years include more drainage and reseeding; a switch from hay to silage; a move away from dicotyledonous fodder crops and barley grown for fodder to intensive grass and forage maize; increased stocking rates; and the introduction of avermectins. In the lowlands, there was a trend to larger, specialised farms with large fields and fewer hedgerows and increased use of pesticides and fertilisers.

2.1 Land Drainage

Land drainage grants were introduced in 1940 and abolished in 1987. The area of land drained reached a peak of 100 000 ha year[-1] in the 1970s. Most land drainage was undertaken to improve extensively managed or rough grassland but 40% was carried out for conversion to arable cropping. Remaining wet grassland habitats have also been affected by the drainage of adjacent arable land.[7] Wet grasslands support distinctive plant communities and often contain rare species that are adapted to local conditions. Land drainage has also impacted upon wetland bird species, such as waders. Four bird species in particular have been affected: lapwing (*Vanellus vanellus*), redshank (*Tringa tetanus*), curlew (*Numenius arquata*) and snipe (*Gallinago gallinago*). Other species such as the starling (*Sturnus vulgaris*) have also been affected, as drier land alters the availability of soil-surface invertebrates.[8] The decline of the water vole is probably partially due to widespread drainage and canalisation of rivers.[9]

Drainage is often the first step in the process of improvement, followed by intensification of farming on drained land, including reseeding, ploughing and fertiliser application. This intensification has had subsequent impacts on bio-diversity. More rapid and denser grass growth of competitive species as well as higher stocking densities may occur in the case of managed grassland. These changes alter the habitat properties significantly, reducing both grazing oppor-tunities for wildfowl and the range of seed resources available for use by granivorous birds.[10] Drainage of wetlands can also reduce soil penetrability for probing birds and access to the soil surface may be reduced by more vigorous spring plant growth, reducing access for breeding birds.[11] Declines in bird species as a result of land drainage have occurred in both the lowlands[12] and the uplands.[13]

Although land drainage may have negative impacts upon biodiversity, particularly for rare wetland communities, ditches and drains themselves can provide a rich habitat within arable landscapes and can be final refuges for species that have declined due to field drainage.

2.2 Decline of Mixed Farming and Changes in Crop Rotations

The polarisation of farming, with arable farming now dominating in the east and grassland in the west, has reduced the diversity of habitats and resources associated with mixed farming. Agricultural intensification has brought about a change in the range and rotation of crops being grown. Management practices, vegetation structure and duration to harvest have altered. This has potentially had an impact upon biodiversity; however, more research is required to understand the scale at which changes are likely to have had a negative impact upon populations. A loss of crop diversity is often cited as one of the key factors in the decline of the brown hare (*Lepus europeaus*),[14,15] although hares are still relatively common on arable farms.[16]

The separation of pastoral and arable farming systems has led to declines in bird populations, and grassland management in arable landscapes would improve habitat diversity within farmland.[17] Equally, it is also suggested that the presence of arable habitat within grassland landscapes can be vital to the survival of key granivorous species such as grey partridge (*Perdix perdix*), skylark (*Alauda arvensis*) and corn bunting (*Miliaria calandra*) in grassland regions, in order to prevent local extinctions.[18]

The increasing availability of a large variety of pesticides and fertilisers, combined with the rising price of cereals, has led to grass leys (*i.e.* temporary grassland), once an important part of the arable rotation, being less popular as a means of controlling weeds and insect pests or for maintaining soil fertility. The introduction of fungicides has allowed many farmers to dispense with "break crops" such as grass or roots, previously used to control cereal fungal diseases. As a consequence of these changes, some arable weeds have declined, whereas others, such as barren brome (*Anisantha sterilis*), have increased. Weed species composition and diversity within different crop types will depend upon the management, *e.g.* fertiliser, pesticide and harvesting/cutting requirements of that particular crop type. For example, oilseed rape often has a higher level of broadleaved weeds than cereals, because they are more expensive to control and the yield benefit does not justify the additional herbicide cost. Sugar beet is sown in late spring so is more likely to contain spring-germinating species such as fat hen (*Chenopodium album*), a key bird food species.[19]

Thus, reduced habitat heterogeneity in all types of farming landscape is likely to have been an important factor in biodiversity losses over the last forty years as mixed farming has declined. However, mixed rotations, including arable crops and grassland grazed by livestock, are still key components of most organic farming systems.

Other changes in cropping practices have also had detrimental effects on biodiversity. For example, the decline in cultivation of fodder crops such as turnips in favour of field beans and, more recently, the increase in the growth of forage maize at the expense of barley in southern and western England have not been beneficial to bird populations. Changes in cropping practices may not always have a negative impact upon all species, however. For example, wood-pigeons (*Columba palumbus*) declined as clover was no longer sown on winter stubble during the 1960s, but increased in numbers when an alternative winter food source became available: the young leaves of autumn sown oil-seed rape.[8]

2.3   Fertiliser

The use of inorganic nitrogen fertiliser has doubled during the second half of the twentieth century.[20] This has had important implications for biodiversity, particularly for plant species and associated fauna. Increased fertiliser use, often in conjunction with reseeding to competitive ryegrass (*Lolium perenne*), has resulted in major losses of botanical diversity in the majority of UK grasslands.[1] One of the most important factors affecting plant diversity is nutrient availability and thus the productivity of a habitat. Since the 1940s inorganic fertilisers have increasingly been used, which allow greater concentrations of nutrients to be applied with a quicker release time into the soil, though reliance on inorganic fertilisers reduces the amount of organic matter in the soil and may affect soil chemistry.[4] Inorganic fertilisers promote rapid growth of a few competitive species, reducing light levels within the crop and preventing the growth of other plants.[21] Along boundaries and hedgerows, misplaced nitrogen fertiliser can alter the nutrient balance, encouraging the growth of nitrophilous annual weeds such as cleavers (*Galium aparine*) and common nettle (*Urtica dioica*).[22]

The stimulation of grass growth by fertiliser in grasslands renders the sward unsuitable for ground-nesting birds, and eliminates many broad-leaved plant species through competition.[8] Loss of plant species has indirect effects on birds too, as insect diversity and abundance is reduced. Rapid growth in spring stimulated by nitrogen allows early and more frequent cutting, so that birds do not have time to complete breeding before cutting destroys nests.

Other nutrients, such as phosphorus and potassium, can also lead to changes in plant species in treated fields. In trials conducted in a species-rich hay meadow on a Somerset peat moor, phosphorus was more important than nitrogen in determining both biomass production and plant species change.[23] Grazing levels after cutting, along with fertiliser inputs, also contributed to botanical change. Liming, in order to reduce the acidity of soil before cultivation, has probably contributed to the decline of corn spurrey (*Spergula arvensis* L.) and corn marigold (*Chrysanthemum segetum*).[24]

2.4 Pesticides

Highly toxic organochlorine pesticides used during the 1960s and 1970s have now been withdrawn following the well-documented decline in biodiversity that resulted from direct impacts of their use.[25] Screening of pesticides for toxicity to non-target organisms combined with risk assessment has reduced risks to such species; for example, the number of poisoning incidents involving wild mammals and pesticides has decreased over the last 10 years, from 20 in 1996 to 12 in 2005. In 2005 only two mammal deaths were caused by the approved use of pesticides, rodenticides in both instances.[26] This reflects the generally lower toxicity of modern products, improved education and tighter controls governing their use.

Today, indirect effects of pesticides (*i.e.* effects operating through the food chain) are of greatest concern for vertebrates; in particular, from broad-spectrum herbicides and insecticides. Broad-spectrum pesticides, insecticides especially, have a major long-term effect on non-target invertebrates,[27] many of which are important as biological pest-control agents or as important links in the food chain of farmland faunal groups.[28] In particular, the application of broad-spectrum pesticides in agricultural systems has decreased bee populations dramatically.[29]

Herbicides impact upon bird populations by either (1) reducing the abundance of, or eliminating, non-crop plants hosting arthropod foods for birds, particularly during the breeding season, or (2) depleting or eliminating weed species, which provide food for herbivorous or granivorous species.[30] Herbicides reduce butterfly abundance indirectly when they cause the loss of larval food-plants and nectar or pollen sources for adults, particularly in boundary vegetation.

In addition to the toxicity of the pesticide itself, the timing of pesticide application, the indirect effects of the pesticide and impacts on non-target species are all important to invertebrate survival. Pesticides may reduce invertebrate abundance through direct toxicity, but also indirectly by restricting food supply or altering habitat. Pesticides are now routinely screened for side effects against non-target invertebrates for the purpose of registration.

The direct impact of pesticides on different invertebrate groups is highly dependent upon the timing of application. For example, autumn applications tend to drift into field boundaries more because crop and marginal vegetation heights are lower[31] and may therefore contact invertebrates over a larger area.

The importance of indirect effects of pesticides on birds was first identified for the grey partridge.[32,33] Pesticides (insecticides and herbicides) were shown to be important in limiting chick survival by reducing the supply of invertebrates important in chick diet, and the implementation of conservation headlands, whereby invertebrate densities were increased by restricting use of these pesticides at the edges of cereal crops, was shown to increase chick survival.[34,35] In a review of indirect effects of pesticides,[36] grey partridge was considered to be the only species for which such effects had been conclusively demonstrated, though there was circumstantial evidence for a number of other species. Corn bunting brood condition and the probability of nest survival were both correlated with the abundance of insect food close to the nest and, furthermore, the abundance of chick-food was negatively correlated with the number of insecticide applications.[37] More recent studies have shown similar results for yellowhammer (*Emberiza citrinella*).[30,38,39]

The exposure of mammals to pesticides can be very variable, depending on how the pesticide is applied and the feeding behaviour of the mammal species,[40] and though the short-term impacts may be measurable, long-term effects are unclear.[41] Species that frequent arable land, such as badgers (*Meles meles*), bank and field voles (*Clethrionomys glareolus* and *Microtus agrestis*) and deer are likely to suffer exposure to agrochemicals.[42]

Use of pesticides continues to increase in intensity: the area of arable crops increased by 3% between 1994 and 2004, but the pesticide-treated area increased by 42%, though the weight of pesticides applied had decreased by 4%.[43] This increase in area treated despite a decrease in area grown reflects an increase in average number of sprays applied to each crop, plus an increase in the number of products used. The average number of pesticide sprays applied to a crop has increased from four in 1994 to over five in 2004. Also the number of products used increased from an average of seven products per crop in 1994 to almost eleven in 2004. On grasslands, pesticides are used less intensively, though herbicides may be used to control some broad-leaved weeds in order to maximise grass quality for livestock grazing.

Arable weed seedbanks have been significantly depleted by intensive herbicide usage over the last fifty years, so that some plants once termed "weeds" are now species of conservation concern in the UK.[44] Chickweeds (*Stellaria*), knotgrass and persicarias (*Polygonum*) and goosefoots (*Chenopodium*) are important food sources in the autumn and winter for a large number of granivorous bird species on farmland, such as linnet (*Carduelis cannabina*) and tree sparrow (*Passer montanus*).[19,45] These particular weeds are generally less competitive in winter cereals; thus more selective herbicide use could help to maintain these species. However, between 1970 and 1995 the spectrum of activity of herbicides on weed taxa increased from an average of 22 to 38 taxa.[46]

The impacts of the increased use and efficiency of herbicides are exacerbated by improved seed-cleaning techniques and the development of increasingly competitive, nitrogen-responsive crops, acting to minimise weeds and reduce availability of seed foods. The timing of herbicide applications may also be important; spraying with herbicide in the spring/summer had the greatest impact on plant populations, particularly in terms of a decrease in broad-leaved weed occurrence.[46] During recent decades herbicide use has increased and switched gradually from spring/summer applications to autumn/winter applications, reflecting the trend towards autumn-sown crops over the last 30 years, though follow-up spring applications are often also made.

Pesticides are likely to be one of the most important factors influencing gross levels of abundance of farmland plants.[47] Many of the plant species that remain common on farmland either have prolific and persistent seedbanks or are resistant to or difficult to target with herbicides.[4]

2.5   Field Size and Hedgerows

Increased mechanisation and the demand for productivity in the latter twentieth century led to increased field sizes and subsequent hedgerow loss. Hedgerows were removed to allow the use of large machinery in arable cropping and because their role in stock control became redundant with the loss of livestock from arable farms.[48]

Hedgerows and field margins are important habitats for a range of wildlife.[49,50] Thus the removal of hedgerows is likely to have had a significant impact upon biodiversity. Hedgerows both facilitate and restrict invertebrate movements and flight activity.[51] For example, butterflies disperse along linear features such as hedgerows, but may also perceive them as barriers to movement.[52] Hedgerows can also restrict the movement of beetles, causing aggregations around field boundaries.[53] It has been shown that hedgerows are particularly associated with the presence of bee species.[54]

Hedgerows are important habitats for a number of species of bird, including yellowhammer.[55] Hedgerow removal or abandonment of hedge management, excessive cutting of hedgerows, incorrect timing of hedgerow cutting, filling or clearing of ditches and other intensification that has impacted upon field margins such as cropping or grazing right up to the field edge, have had a long-term impact on hedgerow bird species.[55]

Hedgerows are important mammal habitats in the agricultural landscape, as a source of food and shelter and as corridors of movement between other habitats; over 20 mammal species are known to live or feed in hedgerows. Hedgerows are also important to predators and can provide one of the most prey-rich habitats for species like the weasel.[56]

Cutting or flailing during the autumn, when fruit is setting, or in the winter can lead to the loss of important food resources for small mammals. Reduction of hedgerow size due to an intensification of management and the resulting loss of food and cover has been implicated in the decline of the common dormouse (*Muscardinus avellanarius*).[57]

Large scale losses of hedgerows have occurred as a result of agricultural intensification and, although this loss appears to have been halted and new plantings undertaken, the ecological value of these new hedgerows, particularly compared to ancient hedgerows that have been lost, has yet to be assessed.[48]

2.6 Autumn Sowing

Winter cropping has increased at the expense of spring-sown crops over the past 40 years due to the availability of higher-yielding cereal varieties capable of overwintering. The percentage of spring-sown wheat and barley declined from over 70% in 1968 to less than 20% in 1998.[58,59] Key differences in associated flora and fauna occur between winter-sown crops (cereals and oilseed rape) and spring-sown crops (potatoes, sugar beet and maize); for example, spring root crops have lower abundances of beetles and spiders.[22,60]

The increase in autumn tillage has changed the composition of weed communities and reduced stubble feeding grounds over winter.[5] The reduction in the presence of winter-stubble is highly significant, as areas of winter-stubble positively influence national trends in breeding and population recovery of key farmland bird species such as skylark and yellowhammer.[61] Winter cropping has resulted in the loss of breeding habitat for ground-nesting birds such as lapwing[62] and skylark,[63] which prefer the more open structure of spring-sown crops.

Tillage, particularly ploughing, can cause high mortality in populations of soil dwelling Diptera (two-winged flies) larvae.[64] Ploughing also has a strong negative impact on spiders,[65] with autumn cultivation being damaging,[66] and is also reported to be detrimental to moths[67] and Hemiptera.[68] Symphyta (sawfly) abundance has declined rapidly due to increase in autumn cultivation, which destroys over-wintering larvae.[69]

An increase in area of winter-sown cereals may have had an adverse impact on the harvest mouse (*Micromys minutus*). Modern cereals grow and ripen quickly and so are ready to harvest in late summer, which corresponds to the peak breeding season of the harvest mouse. The loss of nests and young in cereals, combined with a reduction in rough grassy areas which can also provide suitable nesting habitat, has probably contributed to the decline of the harvest mouse on farmland.[70]

A change to winter cereals from spring cereals leads to a reduction in weed density and species diversity and, in the UK, a change to autumn cultivation may have contributed to the decline of spring-germinating species such as cornflower (*Centaurea cyanus*), corn marigold and red hemp-nettle (*Galeopsis angustifolia*).

Earlier harvesting of crops may limit the breeding season of late nesting bird species, for example, corn bunting.[71] Modern, efficient combine harvesters leave less wastage and grain remaining in stubbles after harvest, which, when combined with the effects of more efficient weed control, further reduces the availability of food for seed-eating birds and other seed predators over the autumn and winter period.[1]

2.7　Management of Grassland

Grassland management can have major impacts upon feeding and nesting bird population dynamics within the field in the UK. Intensification of grassland management has led to increased vegetation density and homogenisation. Longer vegetation in grassland systems can enhance food supplies for several farmland bird species (*e.g.* Tipulid larvae).[72] Conversely, sparse vegetation benefits aerial hunters such as kestrels (*Falco tinnunculus*).[73] Of the 20 species included in the Farmland Bird Index, it has been suggested that 15 of these species are likely to benefit from shorter swards for foraging and predator detection, for at least part of the year.[74]

For some species (*e.g.* lapwing, starling, barn owl (*Tyto alba*)) mosaics of short and long vegetation may provide the optimum conditions.[74] It is suggested that ideally these mosaics should be varied temporally as well as spatially (managed as short-term leys mixed with permanent pasture), with low-intensity cattle grazing over the autumn and winter, to maximise the range of bird habitat available.[75] Similar conclusions were reached in a modelling study of coastal grazing marshes, indicating that the heterogeneity of grass sward height was more important in determining species presence than mean sward height. Complexity of the grass sward and surface topography favoured the presence of ground nesting birds.[76]

Agricultural improvement of grass swards has resulted in higher yields, allowing more frequent cutting and higher grazing densities. The conversion of rough grazing, the switch from hay to silage and the loss of temporary grassland in rotation may have impacted upon birds by reducing insect food supplies.[77] Grassland intensification is detrimental to butterfly populations.

Ploughing and reseeding old pastures with rye-grass (*Lolium*) swards eliminates all known larval food plants of British butterflies,[78] and pasture improvement coupled with high stocking levels destroys the structural and botanical diversity of swards, which is needed to support a high diversity of butterfly species.[79]

In upland regions, many areas of grassland have been ploughed and reseeded with high-yielding grass species in order to improve fodder quality. However, these grasslands are generally dominated by perennial rye-grass, which can withstand frequent defoliation and disturbance and responds to high levels of nitrogen use.[10] Plant species diversity is therefore low, as few broad-leaved forbs can survive within these swards. The ploughing and reseeding of unimproved pasture combined with the input of inorganic fertiliser have caused a massive loss of botanical diversity in grassland.[80]

2.7.1   Cutting Management.   Since the 1970s there has been a trend towards silage rather than hay for winter stock feeding. Grassland grown for silage usually consists of highly fertilised reseeded swards. Silage has higher moisture content than hay and so mowing for silage can begin as early as mid-March and can be repeated every few weeks until the autumn. This allows little time for both grass and forb species to flower and set seed and little opportunity for seed to enter the seed bank.[10] In contrast, hay is usually cut in mid to late summer, allowing many plants species to set seed. Moreover, seed is released during handling and transportation and dung from hay-fed stock also contains large quantities of seed, which may be deposited back onto grassland with a possibility of germination. Because of earlier and more frequent cutting, seed set is largely prevented. Sources of food for birds and other animals are thus reduced. Silage cutting, with fast cutting machines, occurs at a time of year when leverets (young hares) are using the grass fields for feeding and cover, resulting in potentially high mortality levels.[81]

2.7.2   Grazing Management.   Uplands and lowlands have undergone a large increase in sheep stocking densities in recent years, which has clearly impacted upon the habitat structure, sward height and composition of grasslands. The number of sheep in the UK increased by 3% between 1984 and 2005 whereas the number of cattle declined by 20% over the same period, though the rate of increase for sheep would probably have been higher and the decline less for cattle without the Foot and Mouth Disease outbreak in 2001. The effects of grazing upon biodiversity are complex and depend upon the type of stock, stocking rate and timing of grazing.[1]

Livestock type can alter the structure and species composition of the sward. Sheep tend to bite off vegetation close to the ground; often selecting plants low in the grassland profile and so produce very short swards. They can be selective feeders, avoiding tall plants and tussocky areas and often selecting flowers over grass stems. Wethers (castrated rams) are less selective grazers and have a lower mineral requirement than lambs or ewes and so will often feed on coarser and less palatable vegetation.[82] Cattle, however, are less selective feeders than sheep and cannot graze as close to the ground so that longer swards are maintained.[20] Also, cattle avoid grazing close to dung-pats, so the longer grass around these results in a more structurally diverse sward.

Timing of grazing can also be important. In one study, species richness of a lowland neutral sward was increased by sheep grazing in spring or winter, provided that spring grazing was followed by only light summer grazing,[83] whereas in another, species diversity, richness and original species composition in a meadow were maintained by autumn and spring grazing.[84] Grazing in spring and summer can be useful in the control of scrub or undesirable weeds but can be detrimental to early flowering plants such as the fritillary (*Fritillaria meleagris*). In the autumn, most plant species have finished flowering and set seed, and winter grazing probably has little effect on most grassland herbs, though poaching (*i.e.* breaking of the sward by livestock trampling, particularly a problem under wet conditions) is possible under heavy grazing, and the low nutritional quality of the grass may require the supplementary feeding of stock.[20]

Grazing acts upon plant communities through defoliation, trampling, deposition of dung and urine and poaching, all of which can alter the relative abundance of species and their competitive abilities. Grazing by animals usually results in greater plant species richness; the sward is kept open, allowing the establishment of forb species in the gaps.[10] However, too little or too much grazing both can have an adverse effect on species diversity. Low grazing levels can lead to patches of tall rank vegetation and deep litter layers, both of which hinder establishment of forbs, whereas heavy grazing produces short, dense swards with little production of seed.

Reductions in pastoral bird populations have been linked to increases in stocking densities of sheep.[85] High stocking densities have reduced seed and insect availability to birds through reduced sward height and plant diversity (such as loss of heather (*Calluna vulgaris*) moorland), impacting upon almost all species of open upland habitats.[8] The value of the sward as nesting and wintering habitat has deteriorated under these conditions and short, uniform swards are poor for shelter and protection from predators.[2,20] Direct impacts of higher stocking densities include increased defoliation intensity and trampling, which destroys bird nests and young.[20] Key species affected include ground-nesting birds such as black grouse (*Tetrao tetrix*) and red grouse (*Lagopus lagopus*).[8] A recent study of lowland grasslands indicates that such intensification may only affect specific insectivorous birds during the summer (for example, buntings, skylark, whinchat (*Saxicola rubetra*) and red-backed shrike (*Lanius collurio*)) and intensification may in fact be beneficial to several species in winter (such as carrion crow (*Corvus corone*) and jackdaw (*Corvus monedula*)) as soil invertebrates are increased.[86] Studies also show that more intense grazing of moorland favours skylarks and the fragmentation of heather favours meadow pipits (*Anthus pratensis*).[87]

Field voles, which have declined over the last 40 years,[4] are very dependent on rough grassland, though if this habitat is not grazed it can revert to unsuitable scrub, and if it is grazed too much, the sward can become too short to attract voles.[88] The loss of unimproved and rough grassland and heavy grazing has also adversely affected pygmy shrews (*Sorex minutus*) and common shrews (*Sorex araneus*),[88] and unimproved and ungrazed pasture has been found to be strongly associated with high numbers of brown hares.[89,90]

2.8   Heather Burning

In uplands, burning of heather has become widely practised as a method of promoting fresh growth. Where red grouse shooting is common, burning is on a small scale in rotation to create a heather mosaic; however, where sheep are dominant burning is large scale and more frequent.

When sensitively managed, burning of heather helps to remove the accumulation of older, woody areas of the plant and also helps to stimulate seed germination and shoot regeneration. These shoots provide a good food source not only for sheep grazed in upland areas, but also for grouse. However, repeated burning can lead to a loss of moorland flora and wildlife habitat, fire-resistant species such as bracken (*Pteridium aquilinum*) can be favoured and, when burning is undertaken in conjunction with heavy grazing, grass species such as purple moor-grass (*Molinia caerulea*) may replace heather.[91] Only a few mammal species such as the red deer (*Cervus elaphus*) and the mountain hare (*Lepus timidus*) are primarily associated with upland habitats and only the latter appears to be affected by the practice of burning to encourage heather regeneration. Sheep grazing can also result in the loss of heather and the spread of course grasses and the sheep themselves compete directly with the hares for food and cause disturbance to feeding hares.[92]

2.9   Grain Storage and Animal Housing

Keeping animals inside yards and/or sheds and off pasture fields can have impacts on biodiversity, as grazing may improve floral and faunal diversity both in terms of species and structure. For example, it has been found that keeping cattle inside can result in fewer ground and dung beetles present as prey for bat species.[92] Keeping animals indoors during the winter can reduce food sources for birds that feed on the grains provided for the livestock, for example, corn buntings. Also, the elimination of rickyards and other sources of grain around farmyards in mixed farmland has reduced the availability of food for corn buntings, with the highest impact during the breeding season,[93] and has removed an easy source of rodent prey for barn owls.[94] Tidiness and hygiene around farmyards and "bird-proof" storage facilities also limit food availability for birds; it is believed such measures may be responsible for the recent decline in the rural house sparrow (*Passer domesticus*) population.[95]

Another factor leading to a reduction in grain availability in the field is the efficiency of modern combine harvesters, producing less waste and resulting in lower densities of grain left in the stubble after harvest. In conjunction with efficient weed control practices and the effects of keeping animals indoors (see above), food availability for seed-eating birds and other seed predators is much reduced over autumn and winter.

2.10 Veterinary Medicines

Veterinary medicines are widely used to treat disease and protect the health of animals. Release of veterinary medicines to the environment occurs both directly (for example, the use of medicines in fish farms) and indirectly, via the application of animal manure containing excreted products to land. There is a need to explore the links between fertiliser inputs, predation rates, anti-helminithic (worming agents) treatments for livestock and bird habitat suitability in grasslands,[20] and also the implications of veterinary medicine use for biodiversity in uplands.

Pasture invertebrate assemblages are potentially threatened by modern livestock endo-parasite control practices. The best-studied group of endo-parasitic treatments, in terms of potential environmental impacts, are the avermectins, which have been widely used for the past twenty years. While these chemicals offer very effective endo-parasite control, they do not decompose well and remain active in cattle dung at least five weeks after treatment.[96] Avermectin residues are excreted in the faeces of treated animals and, being insecticidal, may reduce the numbers and diversity of invertebrates associated with dung, many of which are important prey items for birds.[20] A large number of birds feed on insects within cow dung (for example starlings, rooks (*Corvus frugilegus*) and jackdaws) particularly in winter when other food sources are scarce and in spring when beetles emerge (coinciding with nesting).[97] Direct effects of avermectins on birds are not evident; however, indirect effects may be caused by a reduction in dung insects, which would have greatest impact at critical times of year, such as during the breeding season or when chicks begin to forage. In addition to direct mortality, avermectins cause non-lethal effects such as reduced invertebrate fecundity. This could depress sensitive populations of dung beetles such as *Aphodius* spp.[98]

Dung, particularly that from cattle, is an important source of invertebrate prey for several bat species.[99] For bat species such as the lesser horseshoe (*Rhinolophus hipposideros*) and Natterer's (*Myotis nattereri*), dung-associated dipterans are an important food source and, for larger species such as the serotine (*Eptesicus serotinus*), dung beetles are particularly important in the diet in late summer and autumn when the young bats are preparing for hibernation.[99] A shortage of suitable prey could have serious consequences for these bat species.

2.11 Supplementary Feeding

Supplementary feeding of livestock, particularly of sheep on upland areas, is often carried out over the winter months when the nutritional value of the vegetation is low. In such instances, instead of spreading their grazing over an area, flocks tend to concentrate around the sites where supplementary feed is put out, leading to a degradation of the upland heather vegetation.[91,94] Hay is often placed upon areas of old heather to prevent it from blowing away, but this also leads to a concentration of trampling and grazing on vegetation least able to withstand it. Urea-based feed blocks can also stimulate the sheep to eat more roughage, which is usually taken as heather,[91] and this can also contribute to the decline of the vegetation. Ideally, supplementary feeding on heather moorlands should be given on areas of coarse grass or dead bracken, away from heather stands.

3 Recent Changes in Agricultural Practices

Since the zenith of high productivity farming in the 1980s, a number of changes have occurred, which have been largely driven by successive reforms of the European Common Agricultural Policy (CAP). Accumulation of surpluses plus a need to reduce burgeoning expenditure led to a reduction in incentives to maximise output per unit area, coupled with falls in prices for many agricultural commodities. This led to changes in production techniques designed to increase efficiency and reduce costs, with the aim of optimising, rather than maximising, output and an increased emphasis on sustainability of production. At the same time, interest has grown in alternative enterprises to supplement the traditional crop and livestock products of agriculture.

In addition, a growing realisation of the impacts of intensification upon biodiversity and other environmental attributes led to a gradual but increasing emphasis in agricultural policy upon environmental protection and biodiversity conservation, including the introduction and development of a number of "agri-environment" schemes, which provide payments for environmentally beneficial management. In parallel to these developments, a growing public awareness of the negative effects of intensive production and concern about the health implications of production methods has increased the market for food produced from agricultural systems perceived to be environmentally benign, such as organic farming.

3.1 Farming Systems

Although organic farming has been practised for several decades, interest has grown in recent years because of the perceived benefits for human health and the environment. In recognition of this, Government incentives for conversion to organic farming were introduced. In England, The Organic Aid scheme was launched in 1994 and superseded by the Organic Farming Scheme in 1999. This has now closed, to be replaced by the Organic Entry Level Scheme. Equivalent schemes in other parts of the UK are the Organic Aid Scheme in Scotland and Organic Farming Schemes in Wales and Northern Ireland.

Concerns about the sustainability of intensive production methods also led to an increased interest in the development of "integrated" farming systems, which continue to utilise pesticides and inorganic fertilisers that are largely avoided by organic farmers, but aim to minimise the use of these inputs through integration with cultural control of weeds, pests and diseases, avoidance of nutrient losses and improved efficiency of nutrient use. Guidelines for Integrated Crop Production were drawn up in 1993,[100] based on the IOBC definition of integrated farming which can be summarised as "a farming system that produces high quality food and other products by using natural resources and regulating mechanisms to replace polluting inputs and to secure sustainable farming".

Early studies comparing the effects on biodiversity of organic and conventional farming were often inconclusive, but in recent years evidence for biodiversity benefits from organic farming has grown. It is hard to define exact elements of organic farming that give rise to specific benefits; however, in general, the combination of mixed crop and livestock farming, varied crops, low chemical pesticide and inorganic fertiliser use, smaller fields and a larger proportion of field boundaries, which is characteristic of organic systems, tends to improve their suitability as habitat, particularly for birds.[101,102] There is a general consensus that greater abundance and/or species richness of birds is found on organic rather than conventional farms, particularly for skylark, blackbird (*Turdus merula*) and greenfinch (*Carduelis chloris*) populations.[102]

Total bat activity and foraging activity both were found to be significantly higher on organic farms[103] and the activity levels of small mammals were found to be higher in organic than in conventional fields.[103] In both studies, increased food abundance as a result of sympathetic management, particularly of hedgerows, was cited as the likely factor influencing mammal biodiversity.

Organic farming has been reported to benefit ground beetles (carabids),[104] spiders[65] and butterflies,[105] although other studies found no significant differences between butterfly populations on organic and conventional farms[106] and that species richness of carabids was higher on conventional farms than on organic farms.[107] Other invertebrate orders such as staphylinid beetles have been found to be more abundant in conventional fields.[104] Possible explanations for this are that staphylinids, a group of highly mobile beetles, may avoid pesticide applications more easily than other less mobile groups, that the higher crop density in conventionally managed fields creates a more humid microclimate which suits this group, or that some of the groups with which staphylinids compete are less abundant in conventionally managed fields. The authors of a study which concluded that organic farming can result in a more vigorous carabid fauna in terms of species richness and activity suggested that effects of herbicides/pesticides and different levels of weediness between conventional and organic farming may have accounted for this difference.[108]

Within grassland systems, any differences between organically and conventionally managed pastures were less marked, though it appeared that organic permanent pasture contained more typical grassland species and a greater species richness, especially of forbs.[102] Hedge bottom vegetation also had higher species diversity on organic than on conventional farms. This is likely to be due to the lack of herbicide drift and higher rates of immigration from the greater species pool available in surrounding areas on organic farms.[102]

Integrated Farm Management also promotes more environmentally sensitive approaches to farming (see http://www.leafuk.org) and is perhaps more economically viable for the majority of farmers. Even fewer studies are available that assess the performance of integrated farming in comparison with conventional farming for biodiversity, than for organic farming. There is some evidence that integrated farming systems may benefit seed-eating birds such as chaffinch (*Fringilla coelebs*) and yellowhammer in particular and, in general, bird numbers have been found to increase.[109]

Conservation Grade is an emerging farming system. Farmers have to take 10% of their land out of food production and devote it to habitat creation so that wildlife can be protected. This builds upon Integrated Farm Management practices, providing further benefits for wildlife.

Many researchers advocate mixed farming as key to the success of programmes to sustain farmland bird populations in the UK.[74,110] Heterogeneous landscapes, mixed cropping and crop rotation, grassland within arable landscapes (or arable cultivation within grasslands) and conservation areas are important elements which can often be implemented as options within agri-environment schemes.

Overall, it has been found that other aspects of agricultural practice, such as crop type and the location of the farms assessed in the studies, tend to be more significant than the differences in the farming practices themselves.[111] Individual farming practices in time and space within each farming system are what really define the benefits to all organisms, including birds, although the structure of the farm defined by the farming system may facilitate key beneficial practices (for example, small fields and more extensive boundaries on typical organic farms).

3.2 Reduced Cultivation Systems

There has been renewed interest recently in reduced forms of cultivation, variously known as minimal cultivation, non-inversion tillage, conservation tillage, eco-tillage and, in its extreme form, as direct drilling or no-till. The common factor is the absence of ploughing (*i.e.* soil inversion), formerly considered essential for weed control and to bring fresh soil to the surface to provide a seedbed. Improved cultivation machinery combined with effective non-selective herbicides (in particular glyphosphate) has made it possible to establish many crops without ploughing, thus providing savings in both time and cultivation costs. Additional benefits are improved soil structure and reduced erosion risk. Conservation tillage is thought to provide better cover for ground-nesting birds compared to conventional tillage or where tillage is used to control weeds.[112] Higher densities of birds and higher productivity by nesting passerines have been observed in some North American studies, but there have been few studies into the effects of this form of tillage on farmland birds in northern Europe to date.[112,113] However, a recent UK study found that gamebirds, skylarks and granivorous passerines all occupied a greater proportion of fields established by non-inversion tillage than by ploughing in the late winter period, though there were no differences early in the winter.[114]

Numbers of earthworms generally increase under reduced tillage systems, but results of studies on arthropods are often conflicting or inconclusive.[112] It appears that different species may respond to cultivations in different ways, but that large species are more vulnerable to cultivations than smaller ones. Minimal cultivations can exacerbate slug problems, thus increasing the need to apply molluscicides. The effects of different forms of soil cultivation and seedbed conditions at the time of sowing upon slugs have been well documented.[115] The more thoroughly the soil is worked, the more effective cultivations become at reducing slug numbers,[116] while direct drilling of land has been reported to give the highest slug numbers.[117]

Many plant species found in arable crops are annuals, which are able to grow and set seed in the time between the sowing of the crop and the post-harvest cultivation. Many of these seeds can remain dormant in the seedbank for several years and species with long-lived seeds such as the common poppy (*Papaver rhoeas*) have persisted on arable land, whereas species like the shepherd's needle (*Scandix pecten-veneris*) that have shorter-lived seeds have declined.[80] The introduction of reduced cultivation systems means that seed is not buried during cultivation and has led to the increased abundance of some perennials and annual grass weed species such as black-grass (*Alopecurus myosuroides*).[112] and may encourage the establishment of species whose seeds are dispersed by wind, such as perennial sow-thistle (*Sonchus arvensis*).[24] However, this system of cultivation may also result in the decline of arable dicotyledons with persistent seedbanks, which benefit from soil inversion.[24]

3.3 Set-aside

Set-aside land was introduced in 1988 as a supply control measure to limit overproduction of cereals and other arable crops. The generation of environmental benefits was not one of the original objectives of set-aside; nevertheless it was hoped that it might improve habitat for farmland birds by encouraging broad-leaved plants and improve arthropod and seed availability. However, natural regeneration may produce dense vegetation with a limited arthropod fauna, depending upon soil type and situation. Even sown grasses may produce a cover of limited value.[22]

In 1992, rotational set-aside was introduced, which led to the creation of more desirable habitats that are more beneficial to farmland biodiversity, depending also upon the seeds that are sown: cereals, brassicas and red clover (*Trifolium pratense*) benefit insects, for example.[118] The most commonly adopted management for rotational set-aside is to allow fields to regenerate naturally a cover of vegetation without any input of agrochemicals.[118] It has been found that a typical, naturally regenerated set-aside field, in the first season of establishment, generally develops a mixture of crop volunteers and opportunistic arable plants with much bare ground. The vegetation on non-rotational set-aside continues to develop into a perennial grass sward after another 1 3 years, due to natural succession following the cessation of soil disturbance. Studies have shown that most species prefer set-aside land as habitat in summer and winter compared to other cropped areas,[119–121] skylark in particular.[122]

Set-aside and fallow land has been associated with higher numbers of hares, probably because these areas create habitat diversity within the agricultural landscape, increasing the variety and amount of food available and providing cover throughout the year.[15,123] The positioning of set-aside may be important; blocks of set-aside were used by wood mice (*Apodemus sylvaticus*) and strips adjacent to margins were avoided, possibly due to the increased risk of predation in the area close to the hedgerow.[124] Management practices, such as sowing with a grass seed mix, mowing and length of time left *in situ*, can all be a positive influence on the attractiveness of the area to small mammals like the field vole.[125]

In addition to natural regeneration and sown grass swards, set-aside can also be sown to "wild bird cover" (defined as an unharvestable mix of at least two crop groups, *e.g.* cereals and brassicas) and non-food crops. Wild bird cover can be a valuable source of food for seed-eating birds in winter.[126,127] Non-food crops are usually grown to produce oil for industrial use (*e.g.* oilseed rape, linseed) or energy crops (see below). Where non-food crops are of the same type as those grown for food, production systems, and hence impacts on biodiversity, are essentially the same.

3.4 Energy Crops

A number of novel crops have been investigated as potential alternatives to those currently grown. Some have achieved commercial success, *e.g.* borage, but so far these have remained "niche" crops, grown on a very limited area. However, one class of novel crops has the potential to cover significant areas of farmland in the near future, *viz.* crops grown as a source of renewable energy, in order to reduce reliance on fossil fuels and help meet targets to reduce greenhouse gases. Thus, substantial increases in the area of non-food "energy crops", such as *Miscanthus*, short-rotation coppice and oilseed rape, are likely in the near future.

Annual arable crops such as wheat, barley, potatoes and sugar beet, which are all used to produce bio-ethanol, and oilseed rape grown as a source of biodiesel, are likely to contain the same invertebrate communities as conventional crops.[128] A number of studies have found that oilseed rape is preferred by some bird species to other crops such as wheat and barley.[129] However, any changes in crop varieties or management, particularly in the use of insecticides and herbicides, could lead to changes in biodiversity and insect communities; an increase in these agrochemicals may directly reduce insect populations and the presence of food and cover, whereas a decline in chemical inputs could lead to an increase in arable weeds and greater insect numbers and diversity.[128] Overall, it is suggested that there will be little difference in the environmental impact of growing annual crops for bio-fuel instead of food. There is only likely to be a negative impact if they are grown on current set-aside land.[129]

Perennial energy crops are relatively undisturbed, allowing the development of varied ground vegetation, particularly in the field margins.[130,131] The most common bio-energy crop in the UK is willow short rotation coppice (SRC). A study of the insect species associated with British trees[132] found that five native species of willow in Britain supported 450 phytophagus insects and mites, so there is potential for willow SRC to increase the insect diversity of an area as compared with conventional arable crops. These SRC plantations may be beneficial not only to those invertebrates that live on the trees themselves, but the development of a varied ground flora, particularly in the field margins left unplanted by trees, may also encourage a range of insects. In a four-year project comparing vegetation within willow short rotation coppice with the previous land use of arable farming, the coppice supported a richer plant community in each year of the study.[131] In pre-commercial SRC studies in the UK, overall bird density and species diversity were higher compared to arable crops and managed grassland.[133] This is perhaps because they are broadly similar to traditional coppice woodland that is attractive to many bird species, although the management systems are quite different, resulting in differences in flora and invertebrate communities.[128] In general, willow SRC has been found to contain more bird species than poplar SRC and the more complex structure increases biodiversity.[134] SRC could also provide cover in winter for birds in open areas of farmland and create new areas of suitable habitat for some woodland, scrub and ruderal vegetation species (plants characteristic of highly disturbed ground) in farmland. These areas also provide greater cover and food resources for small mammals such as the wood mouse and are likely to provide a valuable refuge, particularly in the winter when vegetation cover in arable fields provides little protection from predators. However, planting large areas of woody crops on marginal wet grassland in the UK could damage breeding wader populations (such as lapwing, snipe, curlew and golden plover (*Pluvialis apricaria*)) or other birds requiring a more open landscape (such as skylark, yellow wagtail (*Motacilla flava*) and corn bunting).

Perennial grasses also have the potential to be grown as bio-energy crops, but little research on the biodiversity implications has been carried out to date. In one study perennial grass biomass crops were found to support a greater floral diversity than arable fields, due to no or low inputs of agrochemicals,[130] though poor *Miscanthus* growth probably encouraged better weed establishment. *Miscanthus* is, at present, the most widely grown bio-energy grass in the UK, but as it is non-native, it is unlikely to support any specialist invertebrates, though two Lepidopteran species (butterflies and moths) have been identified as possible pest species.[128] The implications of planting *Miscanthus* on bird populations are unknown, as there are no significant areas of crop upon which to base research in the UK.[135] The crop structure is unlikely to be suitable for nesting by open-field ground-nesting birds, except early in the breeding season when the crop is short. It may be suitable for bird species characteristic of tall, rank grassy or ruderal vegetation, such as reed warblers (*Acrocephalus scirpaceus*). Studies of switchgrass crops in North America show that grass energy crops are likely to be suitable breeding habitat for a range of birds.

3.5   Genetically Modified Crops

No genetically modified crops have yet been grown commercially in the UK, but such has been the level of concern at their potential introduction that no discussion of agricultural changes on biodiversity would be complete without some consideration of their potential impact.

Although a wide variety of traits can potentially be introduced into crops through genetic modification, most research to date has been carried out on genetically modified herbicide tolerant (GMHT) crops. A major five-year study known as the 'Farm Scale Evaluations' (FSEs) was carried out from spring 2000 to investigate the potential impacts in comparison with conventional varieties.[136] As part of this research, arable weeds in fields of conventional crops of beet, maize and spring oilseed rape were compared with those under GMHT cropping. Overall, weed diversity in all three crops showed little difference between GMHT crops and conventional crops. Biomass and seed rain were higher in GMHT maize crops but this difference was not detectable in the seed bank. However, the biomass and seed rain were lower in GMHT beet and rape, the seed bank was found to be 20% lower, and it is possible that if this loss was sustained for several years, there could be a detrimental effect on arable weeds in GMHT crop fields.[137,138] Growing a conventional crop within a GMHT rotation could help to replenish seedbanks, though changes in herbicides available and their application timing and rates within conventional crops may also affect arable weeds in the future.[137,138]

The FSEs only compared crops managed in a standard manner. In a study comparing different approaches to managing GMHT crops, it was found that if the management of a GMHT sugar-beet crop was altered by the application of an early overall spray or band spraying, viable seeds in the soil were present at higher levels than in conventional crops, with no loss of yield.[139] Spraying a GMHT crop of fodder beet with glyphosphate later than recommended also had beneficial effects on weed biomass, again without any loss of yield.[140] However, weeds less sensitive to the herbicide glyphosphate could flourish at the expense of other species, with long-term effects on weed populations.

The growth of GMHT crops could lead to an increase in very broad-spectrum herbicides, which will make weed management cheaper and simpler. This might bring environmental benefits in terms of the decreased use of persistent herbicides. Conversely, weed control in GMHT crops may be so efficient that some arable weed species might disappear completely[137] and there may be increased damage through spray drift to adjacent habitats such as hedgerows and ditches.[7]

Among invertebrates, herbivores and their natural enemies, and also pollinators, generally responded in relation to the weed biomass; where this was higher, abundance of both these groups was greater. Detritivores (springtails, Collembola) always occurred at higher densities in GMHT treatments, probably as a result of the later herbicide treatment increasing food resources for this group at the time of assessment.[141]

4 Measures to Benefit Biodiversity on Farmland

4.1 Agri-environment Schemes

Concern over the impact of agriculture on biodiversity and the landscape led to the development of agri-environment schemes, through which farmers were paid to manage land in an environmentally sensitive manner. Environmental measures under the Common Agricultural Policy (CAP) were originally supported under Regulation 797/85, article 19. In the 1992 reform of the CAP, member states were required under the accompanying measures (Regulation 2078/92) to develop agri-environment schemes, with 50% of funding provided by the European Community (75% in Objective 1 areas). Under the Agenda 2000 reform, agri-environment schemes were supported under the Rural Development Regulation 1257/1999 (Chapter VI) and were the only measure under this regulation which member states were compelled to implement, as part of their Rural Development Plan.

The forerunner of agri-environment schemes in the UK was the 1985 Broadland Grazing Marshes Conservation Scheme. This was followed in 1987 by the first tranche of Environmentally Sensitive Areas (ESAs), each supporting specific management practices directed towards the conservation of the wildlife and landscapes characteristic of the area. Eventually, 22 ESAs were established in England, covering some 10% of agricultural land. Environmentally Sensitive Areas were also established in Scotland, Wales and Northern Ireland. However, there was a need for a vehicle to promote environmentally beneficial management outside ESAs, and in 1991 the Countryside Stewardship Scheme was launched. Equivalent schemes were established by the devolved administrations in Scotland (the Countryside Premium Scheme, replaced by the Rural Stewardship Scheme in 2000), Wales (Tir Cymen, succeeded by Tir Gofal) and Northern Ireland (the Countryside Management Scheme). The ESAs were absorbed into single national schemes in Scotland, Wales and Northern Ireland, but were retained in England until the end of 2004, when they were closed to new entrants.

In addition to these major schemes, there were schemes to support organic farming (see Section 3) and a number of smaller schemes limited in scope and timescale with various objectives (*e.g.* the Moorland Scheme, the Habitat Scheme and the pilot Arable Stewardship Scheme, among others).

Agri-environment schemes have also been developed in other European countries, both inside and outside the EU. However, these have varied in their objectives and targets. For example, in Switzerland and the Netherlands, as in the UK, wildlife and habitats have been the priorities, but in Denmark and Germany reduction of agrochemical emissions has been the aim of most schemes, and in France the programme concentrates on the prevention of land abandonment.[142]

Concern about the effectiveness of agri-environment schemes led to a call for more scientific evaluations.[143] A review of the value of schemes in conserving biodiversity concluded that in most of the 62 evaluation studies discovered, the research design was inadequate to assess reliably the effectiveness of the scheme.[142] The authors concluded that there were insufficient scientifically robust evaluations to allow a general judgement of the effectiveness of agri-environment schemes in Europe. In spite of this, supporters of agri-environment schemes point to examples of successes where targeted action has been coupled with effective monitoring.[144–146] Furthermore, at least in the UK, many of the prescriptions adopted are based on research which has already demonstrated positive benefits.[63,126,147,148]

English agri-environment schemes have been through a period of change in recent years.[149] The Policy Commission on the Future or Farming and Food, chaired by Sir Don Curry[150] recommended a new approach, the development of a "broad and shallow" scheme, to run alongside and complement a more demanding "narrow and deep" scheme, similar to the existing Countryside Stewardship. The 2003 reform of the CAP, with an increased allocation of funding to environmental measures and the opportunity to raise additional funds through "modulation" (top-slicing subsidy funding), provided an opportunity to re-structure agri-environment schemes to encourage greater participation and the Curry proposals were translated into the Entry Level and Higher Level of the Environmental Stewardship scheme, launched in 2005. The Entry Level Scheme is open to all farmers and landowners and operates on a points-allocation system: applicants can choose options from a menu, each of which is assigned a number of points per unit area, length, *etc.* All those who reach a threshold number of points are guaranteed entry and payment of a flat rate per hectare of land entered into the scheme. Thus, for the first time, the majority of farmers will be involved in a scheme to encourage positive environmental management.

4.2   Cross-compliance

Cross-compliance, *i.e.* the imposition of conditions on the receipt of subsidies, was first introduced in the 1992 "McSharry" reform of the CAP, in the form of conditions on the management of set-aside to protect environmental features, minimise nutrient losses, *etc.* The UK also applied conditions to livestock headage payments to control overgrazing, but other member states did not implement cross-compliance conditions at that time.

Under the 2003 CAP reform, cross-compliance became mandatory and receipt of payments is now conditional upon compliance with a range of measures. In England, for example, these include a requirement to complete an Environmental Impact Assessment before ploughing permanent pasture, the maintenance of land in "Good Agricultural and Environmental Condition" (GAEC), compliance with Statutory Management Requirements (SMRs) and certain public and animal health measures. Good Agricultural and Environmental Condition standards include measures for soil management and protection and maintenance of habitats and features, including uncultivated land, forestry, Sites of Special Scientific Interest (SSSIs), overgrazing and supplementary feeding, heather and grass burning, uncropped land, stone walls, hedges and watercourses and felling of trees, among others. These provide a baseline standard of environmental management, upon which land managers can build through membership of agri-environment schemes if they so wish.

5   Changing Agricultural Policy and Implications for the Future of Farmland Biodiversity

The most recent round of CAP reform has signalled a fundamental shift in the agricultural support mechanisms. The key element of the reform package was the "decoupling" of subsidies from production, such that the amount received by farmers would no longer be linked to the output from their land. Instead, they would receive an annual single payment, determined either in relation to previous payments (historical model), or in relation to the area farmed (regional model). In reality, options to retain some level of coupling remained and so the extent of decoupling varies between EU member states. The UK and Ireland have opted for a fully decoupled approach (apart from a small element of cattle payments in Scotland), but most other member states have retained some coupled payments.

In addition to the decoupling of subsidies, there was an intention to raise the level of support for environmental measures, as illustrated by the mandatory implementation of cross-compliance conditions and agri-environment schemes in all member states (though most commentators believe that levels of funding for the latter are still lower than optimal).

In England, 90% of payments were historic initially, gradually shifting to 100% regional payments by 2012. "Regions" are Severely Disadvantaged Areas (SDAs), separated into moorland and other SDA land, and all other non-SDA land. The Single Payment Scheme (SPS) replaces 11 previous schemes; however, there are still some anomalies such as the dairy sector, where guaranteed prices and production quotas were still important elements of the support regime initially and reforms were planned to be staged over several years. Scotland and Wales have opted entirely for historic payments, whilst Northern Ireland has a hybrid system.

Further complexities abound, but the underlying principle is that support payments will no longer influence decisions on what and how much to produce from farmland. This increased flexibility may be manifest in a number of ways and will be influenced by a variety of factors. Market prices will have a stronger influence than previously, but farmers will also have the option of not farming the land at all for one or more years, provided it is retained in "Good Agricultural and Environmental Condition". Thus there could be large swings between years, not only in what crops are grown, but in how much of the land is cropped at all. Set-aside was retained as a policy in the reformed CAP and, as the SPS covers a wider range of crop types than the old Arable Area Payments Scheme, the number of farms with a set-aside obligation will be extended. For example, set-aside will now be required in conjunction with land growing horticultural crops such as fruit, vegetables and potatoes, temporary grass less than five years old, and many perennial crops. Management rules are similar to those existing previously, though with increased flexibility in some areas.

What will these changes mean for biodiversity? Much depends on the relative profitability of different enterprises and market forces are notoriously difficult to predict in anything but the short term. A recent analysis[151] indicates that the following changes are likely:

- Sheep will become more common in lowland areas and less common on open moorlands.
- Beef cattle will decrease in numbers in the uplands, but may increase at least temporarily in traditional dairying areas.
- Dairy farms will continue to decline in numbers and those surviving will be larger.
- In more productive arable areas, larger scale block-cropping of wheat and rape with simplified rotations will become more dominant. There will be more fallow but less sugar beet.
- The area of maize will continue to increase.
- In less productive arable areas, some land will grow novel or energy crops, some will fall into disuse, or will be diverted to other land uses or built development.
- There will be continued increase in non-commercial farmland management, *e.g.* "hobby farms", horse grazing and leisure activities.

Outcomes for biodiversity will depend on the relative strengths of these trends and other factors, such as the level of incentives to grow energy crops. Cross-compliance and the Entry Level Scheme (or its equivalent under devolved administrations) should provide environmental benefits, but the extent of the benefits accruing has yet to be tested.

After a long period of relative stability, during which levels of productivity reached hitherto unprecedented levels, agriculture in the UK and elsewhere in Europe is now undergoing a period of rapid change, stimulated by a series of reforms in the CAP support systems since the early 1990s. The countryside of the future could look very different from that existing at the end of the twentieth century. Elements could include, for example, large areas of energy crops, such as short-rotation coppice and *Miscanthus*, areas of unmanaged "fallow" land, areas of intensive crop and milk production, novel crops and new combinations of land uses. Some of the potential developments may threaten biodiversity on farmland, but there are also many opportunities to integrate biodiversity conservation with the agriculture of the future.

Acknowledgements

Most of the material presented here was originally included in a review prepared as part of a project funded by the UK Government Department for Environment, Food and Rural Affairs (Defra), for their Agricultural Change and Environment Observatory Programme.

References

1. N. D. Boatman in "Farming, Forestry and the Natural Heritage: Towards a More Integrated Future", R. Davison and C. Galbraith (eds), Scottish Natural Heritage, Edinburgh, 2006, 39–57.
2. R. J. Fuller and S. J. Gough, *Biol. Conservat.*, 1999, **91**, 73–89.
3. G. C. Daily, "Nature's Services: Societal Dependence on Natural Ecosystems", Island Press, Washington DC, 1997.
4. R. A. Robinson and W. J. Sutherland, *J. Appl. Ecol.*, 2002, **39**, 157–176.
5. D. E. Chamberlain, R. J. Fuller, R. G. H. Bunce, J. C. Duckworth and M. J. Shrubb, *J. Appl. Ecol.*, 2000, **37**, 771–788.
6. P. F. Donald, R. E. Green and M. F. Heath, *P. Roy. Soc. Lond. B Bio.*, 2001, **268**, 25–29.
7. C. Stoate, N. D. Boatman, R. Borralho, C. Rio Carvalho, G. R. de Snoo and P. Eden, *J. Environ. Manage.*, 2001, **63**, 337–365.
8. I. Newton, *Ibis*, 2004, **146**, 579–600.
9. D. W. Macdonald and S. Baker, "The State of Britain's Mammals 2005", Mammals Trust UK, London, 2005.
10. D. I. McCracken and J. R. Tallowin, *Ibis*, 2004, **146 (supplement 2)**, 108–114.
11. M. Ausden, W. J. Sutherland and R. James, *J. Appl. Ecol.*, 2004, **38**, 320–338.
12. R. E. Green and M. Robins, *Biol. Conservat.*, 1993, **66**, 95–106.
13. D. Baines, *Ibis*, 1989, **131**, 497–506.
14. S. C. Tapper and R. F. W. Barnes, *J. Appl. Ecol.*, 1986, **23**, 39–52.
15. R. K. Smith, N. V. Jennings and S. Harris, *Mammal Rev.*, 2005, **35**, 1–24.
16. N. Vaughan, E. -A. Lucas, S. Harris and P. C. L. White, *J. Appl. Ecol.*, 2003, **40**, 163–175.
17. P. W. Atkinson, R. J. Fuller and J. A. Vickery, *Ecography*, 2002, **25**, 446–480.
18. R. A. Robinson, J. D. Wilson and H. Q. P. Crick, *J. Appl. Ecol.*, 2001, **38**, 1059–1069.
19. J. M. Holland, M. A. S. Hutchison, B. Smith and N. J. Aebischer, *Ann. Appl. Biol.*, 2006, **148**, 49–71.

20. J. A. Vickery, J. R. Tallowin, R. E. Feber, E. J. Asteraki, P. J. Atkinson, R. J. Fuller and V. K. Brown, *J. Appl. Ecol.*, 2001, **38**, 647–664.
21. P. J. Wilson in "Aspects of Applied Biology 54, Field margins and buffer zones: ecology, management and policy", N. D. Boatman, D. H. K. Davies, K. Chaney, R. Feber, G. R. deSuoo and T. H. Sparks (eds), 1999, 93–100.
22. J. M. Holland in "Insect and Bird Interactions", H. van Emden and M. Rothschild (eds), Intercept, Andover, 2004, 51–71.
23. F. W. Kirkham, J. O. Mountford and R. J. Wilkins, *J. Appl. Ecol.*, 1996, **33**, 1013–1029.
24. R. J. Froud-Williams, *Weed Res.*, 1981, **21**, 99–109.
25. I. Newton, *J. Anim. Ecol.*, 1995, **64**, 675–696.
26. E. A. Barnett, M. R. Fletcher, K. Hunter and E. A. Sharp, "Pesticide poisoning of animals 2005: investigations of suspected incidents in the United Kingdom", Report of the Environmental Panel of the Advisory Committee on Pesticides, London, 2006.
27. G. P. Vickerman in "Pesticides, Cereal Farming and the Environment: the Boxworth Project", P. Grieg-Smith, G. K. Frampton and A. R. Hardy (eds), HMSO, London, 1992, 82–109.
28. L. Winder, *Ecol. Entomol.*, 1990, **15**, 105–110.
29. J. L. Osborne, I. H. Williams and S. A. Corbet, *Bee World*, 1991, **72**, 99–116.
30. N. D. Boatman, N. W. Brickle, J. D. Hart, J. M. Holland, T. P. Milsom, A. J. Morris, A. W. A. Murray, K. A. Murray and P. A. Robertson, *Ibis*, 2004, **146 (supplement 2)**, 131–143.
31. M. Longley and N. W. Sotherton, *Environ. Toxicol. Chem.*, 1997, **16**, 173–178.
32. G. R. Potts, "The Partridge: Pesticides, Predation and Conservation", Collins, London, 1986.
33. G. R. Potts and N. J. Aebischer in "Bird Population Studies: their Relevance to Conservation and Management", C. M. Perrins, J. -D. Lebreton and G. J. M. Hirons, Oxford University Press, Oxford, 1991, 373–390.
34. M. R. W. Rands, *J. Appl. Ecol.*, 1985, **22**, 49–54.
35. M. R. W. Rands, *Ibis*, 1986, **128**, 57–64.
36. L. H. Campbell, M. I. Avery, P. Donald, A. D. Evans, R. E. Green and J. D. Wilson, "A review of the indirect effects of pesticides on birds", Joint Nature Conservation Committee, Peterborough, 1997, 148.
37. N. W. Brickle, D. G. C. Harper, N. J. Aebischer and S. H. Cockayne, *J. Appl. Ecol.*, 2000, **37**, 742–755.
38. A. J. Morris, R. B. Bradbury and J. D. Wilson, The BCPC conference – Pests & Diseases 2002, Glasgow, 2002.
39. J. D. Hart, T. P. Milsom, G. Fisher, V. Wilkins, S. J. Moreby, A. W. A. Murray and P. A. Robertson, *J. Appl. Ecol.*, 2006, **43**, 81–91.
40. I. Barber, K. A. Tarrant and H. M. Thompson, *Environ. Toxicol. Chem.*, 2003, **22**, 1134–1139.

41. J. A. Skinner, K. A. Lewis, K. S. Bardon, P. Tucker, J. A. Catt and B. J. Chambers, *J. Environ. Manage.*, 1997, **50**, 111–128.
42. R. F. Shore, M. R. Fletcher and L. A. Walker in "Conservation and Conflict. Mammals and Farming in Britain", F. H. Tattersall and W. J. Manley (eds), Westbury Publishing, Otley, 2003, 37–50.
43. D. G. Garthwaite, M. R. Thomas, H. Anderson and H. Stoddart, "Pesticide Usage Survey Report 202: Arable Crops in Great Britain 2004", Defra/SEERAD, 2005, 114.
44. C. Preston, M. Telfer, H. Arnold, P. Carey, J. Cooper, T. Dines, M. Hill, D. Pearman, D. Roy and S. Smart, "The Changing Flora of the UK", The Stationary Office, London, 2002.
45. J. D. Wilson, A. J. Morris, B. E. Arroyo, S. C. Clark and R. B. Bradbury, *Agr. Ecosyst. Environ.*, 1999, **75**, 13–30.
46. J. A. Ewald and N. J. Aebischer, "Pesticide use, avian food resources and bird densities in Sussex", JNCC Report No 296, Joint Nature Conservation Committee, Peterborough, 1999.
47. P. F. Donald, *British Wildlife*, 1998, **9**, 279–289.
48. S. Petit, R. C. Stuart, M. K. Gillespie and C. J. Barr, *J. Environ. Manage.*, 2003, **67**, 229–238.
49. N. D. Boatman and C. Stoate, *British Wildlife*, 1999, **10**, 260–267.
50. C. J. Barr and S. Petit (eds), "Hedgerows of the World: their Ecological Functions in Different Landscapes", IALE (UK), 2001.
51. J. Frouz and M. G. Paolette, *Landscape Urban Plan.*, 2000, **49**, 19–29.
52. J. D. Dover and G. L. A. Fry, *Entomol. Exp. Appl.*, 2001, **100**, 221–233.
53. J. R. Mauramootoo, S. D. Wratten, S. P. Worner and G. L. A. Fry, *Agr. Ecosyst. Environ.*, 1995, **52**, 141–148.
54. P. H. Williams, *Ecol. Entomol.*, 1988, **13**, 223–237.
55. R. B. Bradbury, A. Kyrkos, A. J. Morris, S. C. Clark, A. J. Perkins and J. D. Wilson, *J. Appl. Ecol.*, 2000, **37**, 789–805.
56. D. W. Macdonald, T. E. Tew and I. A. Todd, *Biologia*, 2004, **59**, 235–241.
57. P. Bright and D. MacPherson, "Hedgerow Management, Dormice and Biodiversity", English Nature, Peterborough, UK, 2002.
58. HGCA, "Cereal Statistics 1993", Home Grown Cereals Authority, London, UK, 1994.
59. HGCA, "Cereal Statistics 1998", Home Grown Cereals Authority, London, UK, 1999.
60. J. M. Holland, S. Southway, J. A. Ewald, T. Birkett, M. Begbie, J. Hart, D. Parrott and J. Allcock in "Aspects of Applied Biology 67, Birds and Agriculture", N. D. Boatman, N. Carter, A. D. Evans, P. V. Grice, C. Stoate and J. D. Wilson (eds), 2002, 27–34.
61. S. Gillings, S. E. Newson, D. G. Noble and J. A. Vickery, *P. R. Soc. B*, 2005, **272**, 733–739.
62. R. Sheldon, M. Bolton, S. Gillings and A. M. Wilson, *Ibis*, 2004, **146 (supplement 2)**, 41–49.
63. A. J. Morris, J. M. Holland, B. Smith and N. E. Jones, *Ibis*, 2004, **146 (supplement 2)**, 155–162.

64. J. Frouz, *Agr. Ecosyst. Environ.*, 1999, **74**, 167–186.
65. P. Marc, A. Canard and F. Ysnel, *Agr. Ecosyst. Environ.*, 1999, **74**, 229–273.
66. C. F. G. Thomas and P. C. Jepson, *Entomol. Exp. Appl.*, 1997, **84**, 59–69.
67. R. Fox in "The Changing Wildlife of Great Britain and Ireland", D. L. Hawksworth (ed.), Taylor and Francis, London, UK, 2001, 320–327.
68. P. Kirby, A. J. A. Stewart and M. R. Wilson in "The Changing Wildlife of Great Britain and Ireland", D. L. Hawksworth (ed.), Taylor and Francis, London, UK, 2001, 262–299.
69. N. J. Aebischer, *Funct. Ecol.*, 1990, **4**, 369–373.
70. M. Perrow and A. Jowitt, *British Wildlife*, 1994, **6**, 356–365.
71. N. W. Brickle and D. G. C. Harper, *Bird Study*, 2002, **49**, 219–228.
72. D. I. McCracken, G. N. Foster and A. Kelly, *Appl. Soil Ecol.*, 1995, **2**, 203–213.
73. M. J. Shrubb, *Bird Study*, 1980, **27**, 109–115.
74. M. J. Whittingham and K. L. Evans, *Ibis*, 2004, **146 (supplement 2)**, 210–220.
75. A. J. Perkins, M. J. Whittingham, R. B. Bradbury, J. D. Wilson, A. J. Morris and P. Barnett, *Biol. Conservat.*, 2000, **95**, 279–294.
76. T. P. Milsom, S. D. Langton, C. S. Parkin, S. Peel, J. D. Bishop, J. D. Hart and N. P. Moore, *J. Appl. Ecol.*, 2000, **37**, 706–727.
77. D. E. Chamberlain, *Aspect. Appl. Biol.*, 2002, **67**, 1–10.
78. I. P. Woiwod and A. J. A. Stewart in "Species Dispersal in Agricultural Habitats", R. G. H. Bruce and D. C. Howard (eds), Belhaven Press, London, UK, 1990, 189–202.
79. J. A. Thomas in "The Biology of Butterflies", P. Ackery and R. Vane-Wright (eds), Academic Press, London UK, 1984, 334–353.
80. N. W. Sotherton and M. J. Self in "Ecology and Conservation of Lowland Farmland Birds", N. J. Aebischer, A. D. Evans, P. V. Grice and J. A. Vickery (eds), British Ornithologists' Union, Tring, 2000, 26–35.
81. S. Harris and G. MacLaren, "The Brown Hare in Britain", University of Bristol, 1998.
82. W. C. Shaw, B. D. Wheeler, P. Kirby, P. Phillipson and R. Edmunds, "Literature review of the historical effects of burning and grazing of blanket bog and upland wet heath", English Nature Research Report No. 172, English Nature and Countryside Council for Wales, 1996.
83. J. R. Treweek, T. A. Watt and C. Hambler, *J. Environ. Manage.*, 1997, **50**, 193–210.
84. R. S. Smith and S. P. Rushton, *J. Appl. Ecol.*, 1994, **31**, 13–24.
85. D. J. Pain, D. Hill and D. I. McCracken, *Agr. Ecosyst. Environ.*, 1997, **64**, 19–32.
86. P. W. Atkinson, R. J. Fuller, J. A. Vickery, G. J. Conway, J. R. Tallowin, R. E. N. Smith, K. A. Haysom, T. C. Ings, E. J. Asteraki and V. K. Brown, *J. Appl. Ecol.*, 2005, **42**, 932–942.
87. J. W. Pearce-Higgins and M. C. Grant, *Aspect. Appl. Biol.*, 2002, **67**, 155–163.

88. S. Harris, P. Morris, S. Wray and D. Yalden, "A Review of British Mammals", Joint Nature Conservation Committee, Peterborough, UK, 1995.
89. R. F. W. Barnes, S. C. Tapper and J. Williams, *J. Appl. Ecol.*, 1983, **20**, 179–185.
90. R. K. Smith, N. V. Jennings, A. Robinson and S. Harris, *J. Appl. Ecol.*, 2004, **41**, 1092–1102.
91. P. J. Hudson and D. Newborn, "A Manual of Red Grouse and Moorland Management", Game Conservancy Ltd, Hampshire, UK, 1995.
92. P. A. Morris, "A Red Data Book for Mammals", The Mammal Society, London, UK, 1993.
93. N. W. Brickle and D. G. C. Harper in "Ecology and Management of Lowland Farmland Birds", N. J. Aebischer, A. D. Evans, P. V. Grice and J. A. Vickery (eds), British Ornithologists' Union, Tring, 2000, 156–164.
94. M. J. Shrubb, "Birds, Scythes and Combines", Cambridge University Press, Cambridge, 2003, 371.
95. D. G. Hole, M. J. Whittingham, R. B. Bradbury, G. Q. A. Anderson, P. L. M. Lee, J. D. Wilson and J. R. Krebs, *Nature*, 2002, **418**, 931–932.
96. D. M. Spratt, *Int. J. Parasitol.*, 1997, **27**, 173–180.
97. D. I. McCracken, *Vet. Parasitol.*, 1993, **48b**, 273–280.
98. L. Strong, *Vet. Parasitol.*, 1993, **48**, 3–17.
99. J. Cox, *British Wildlife*, 1999, **11**, 28–36.
100. A. El Titi in "The Ecology of Temperate Cereal Fields", L. G. Firbank, N. Carter, J. F. Darbyshire and G. R. Potts (eds), Blackwell Scientific Publications, Oxford, 1991, 399–411.
101. Anon., "The Effect of Organic Farming Regimes on Breeding and Winter Bird Populations", BTO Research Report 154, Thetford, Norfolk, UK, 1995.
102. D. G. Hole, A. J. Perkins, J. D. Wilson, I. H. Alexander, P. V. Grice and A. D. Evans, *Biol. Conservat.*, 2005, **122**, 113–130.
103. L. P. Wickramasinghe, S. Harris, G. Jones and N. Vaughan, *J. Appl. Ecol.*, 2003, **40**, 984–1007.
104. P. A. Shah, D. R. Brooks, J. E. Ashby, J. N. Perry and I. P. Woiwod, *Agri. Forest Entomol.*, 2003, **5**, 51–60.
105. R. E. Feber, H. Smith and D. W. Macdonald, *J. Appl. Ecol.*, 1996, **33**, 1191–1205.
106. A. C. Weibull, J. Bengtsson and E. Nohlgren, *Ecography*, 2000, **23**, 743–750.
107. A. C. Weibull, O. Ostman and A. Granqvist, *Biodivers. Conservat.*, 2003, **12**, 1335–1355.
108. H. A. Cárcamo, J. K. Niemala and J. T. Spence, *The Canadian Entomologist*, 1995, **127**, 123–140.
109. D. G. Chapple, D. R. Wade, R. M. Laverick and P. J. Eldridge, *Aspect. Appl. Biol.*, 2002, **67**, 129–134.
110. T. G. Benton, J. A. Vickery and J. D. Wilson, *Trends Ecol. Evol.*, 2003, **18**, 179–186.

111. J. M. Holland, S. K. Cook, A. D. Drysdale, M. V. Hewitt, J. Spink and D. B. Turley in "Proceedings of the 1998 Brighton Conference – Pests and Diseases", British Crop Protection Council (ed.), Farnham, 1998, 625–630.

112. J. M. Holland, *Agr. Ecosyst. Environ.*, 2004, **103**, 1–25.

113. H. M. Cunningham, K. Chaney, R. B. Bradbury and A. Wilcox, *Ibis*, 2004, **146 (supplement 2)**, 192–202.

114. H. M. Cunningham, R. B. Bradbury, K. Chaney and A. Wilcox, *Bird Study*, 2005, **52**, 173–179.

115. D. M. Glen, A. M. Spaull, D. J. Mowat, D. B. Green and A. W. Jackson, *Ann. Appl. Biol.*, 1993, **122**, 161–172.

116. P. J. Hunter, *Plant Pathol.*, 1967, **16**, 153–156.

117. J. F. Grant, K. V. Yeargan, B. C. Pass and J. C. Parr, *J. Econ. Entomol.*, 1982, **75**, 822–826.

118. N. W. Sotherton, *Biol. Conservat.*, 1998, **83**, 259–268.

119. D. L. Buckingham, A. D. Evans, T. J. Morris, C. J. Orsman and R. Yaxley, *Bird Study*, 1999, **46**, 157–169.

120. L. G. Firbank, S. M. Smart, J. Crabb, N. R. Critchley, J. W. Fowbert, R. J. Fuller, P. Gladders, D. B. Green, I. G. Henderson and M. O. Hill, *Agr. Ecosyst. Environ.*, 2003, **95**, 73–85.

121. I. G. Henderson and A. D. Evans in "Ecology and Conservation of Lowland Farmland Birds", N. Aebischer, A. D. Evans, P. V. Grice and J. A. Vickery (eds), British Ornithologists Union, Tring, 2000, 69–76.

122. J. A. Vickery and D. L. Buckingham in "The Ecology and Conservation of Skylarks *Alauda arvensis*", P. F. Donald and J. A. Vickery (eds), RSPB, Sandy, 2001, 161–175.

123. M. R. Hutchings and S. Harris, "The Current Status of the Brown Hare (Lepus europaeus) in Britain", Joint Nature Conservation Committee, Peterborough, UK, 1996.

124. F. H. Tattersall, D. W. Macdonald, B. J. Hart, W. J. Manley and R. Feber, *J. Zool.*, 2001, **255**, 487–494.

125. F. H. Tattersall, A. E. Avundo, W. J. Manley, B. J. Hart and D. W. Macdonald, *Biol. Conservat.*, 2000, **96**, 123–128.

126. N. D. Boatman, C. Stoate and P. N. Watts in "Ecology and Conservation of Lowland Farmland Birds", N. J. Aebischer, A. D. Evans, P. V. Grice and J. A. Vickery (eds), British Ornithologists' Union, Tring, 2000, 105–114.

127. N. D. Boatman and C. Stoate, *Aspect. Appl. Biol.*, 2002, **67**, 229–236.

128. G. Q. A. Anderson, L. R. Haskins and S. H. Nelson in "Biomass and Agriculture: Sustainability, Markets and Policies", K. Parris and T. Poincet (eds), OECD, Paris, 2004, 199–128.

129. D. B. Turley, H. McKay and N. D. Boatman, "Environmental impacts of cereal and oilseed rape cropping the UK and assessment of the potential impacts arising from cultivation for liquid biofuel production", Research Review 54, Home Grown Cereals Authority, London, UK, 2005.

130. T. Semere and F. Slater, "The effect of energy grass plantations on biodiversity: Report to the Department of Trade and Industry", 2004.

131. M. Cunningham, J. D. Bishop, H. McKay and R. B. Sage, "Ecology of short rotation coppice: ARBRE monitoring. Report to the Department of Trade and Industry", The Game Conservancy Trust and Central Science Laboratory, 2004.

132. C. E. J. Kennedy and T. R. E. Southwood, *J. Anim. Ecol.*, 1984, **53**, 455–478.

133. R. B. Sage, M. Cunningham and N. D. Boatman, *Ibis*, 2006, **148**, 184–197.

134. R. Sage and P. A. Robertson, *Bird Study*, 1996, **43**, 201–213.

135. D. B. Turley, N. D. Boatman, G. Ceddia, D. Barker and G. Watola, "Liquid biofuels-prospects and potential impacts on UK agriculture, the farmed environment, landscape and rural economy", Report to Defra. Central Science Laboratory, York, 2002.

136. J. N. Perry, P. Rothery, S. J. Clark, M. S. Heard and C. Hawes, *J. Appl. Ecol.*, 2003, **40**, 17–31.

137. M. S. Heard, C. Hawes, G. T. Champion, S. J. Clark, L. G. Firbank, A. J. Haughton, A. M. Parish, J. N. Perry, P. Rothery, D. B. Roy, R. J. Scott, M. P. Skellern, G. R. Squire and M. O. Hill, *Phil. Trans. Roy. Soc. Lond. B*, 2003, **358**, 1833–1846.

138. M. S. Heard, C. Hawes, G. T. Champion, S. J. Clark, L. G. Firbank, A. J. Haughton, A. M. Parish, J. N. Perry, P. Rothery, R. J. Scott, M. P. Skellern, G. R. Squire and M. O. Hill, *Phil. Trans. Roy. Soc. Lond. B*, 2003, **358**, 1819–1832.

139. M. J. May, G. T. Champion, A. J. G. Dewar, A. Qi and J. D. Pidgeon, *P. R. Soc. B*, 2005, **272**, 111–119.

140. B. Strandberg, M. B. Pedersen and N. Elmegaard, *Agr. Ecosyst. Environ.*, 2005, **105**, 243–253.

141. C. Hawes, A. J. Haughton, J. L. Osborne, D. B. Roy, S. J. Clark, J. N. Perry, P. Rothery, D. A. Bohan, D. R. Brooks, G. T. Champion, A. M. Dewar, M. S. Heard, I. P. Woiwod, R. E. Daniels, M. W. Young, A. M. Parish, R. J. Scott, L. G. Firbank and G. R. Squire, *Phil. Trans. Roy. Soc. Lond. B.*, 2003, **358**, 1899–1913.

142. D. Kleijn and W. J. Sutherland, *J. Appl. Ecol.*, 2003, **40**, 947–969.

143. D. Kleijn, F. Berendse, R. Smit and N. Gilisen, *Nature*, 2001, **413**, 723–725.

144. N. J. Aebischer, R. E. Green and A. D. Evans in "Ecology and Conservation of Lowland Farmland Birds", N. J. Aebischer, A. D. Evans, P. V. Grice and J. A. Vickery (eds), British Ornithologists' Union, Tring, 2000, 43–53.

145. E. Knop, D. Kleijn, F. Herzog and B. Schmid, *J. Appl. Ecol.*, 2006, **43**, 120–127.

146. W. J. Peach, L. Lovett, S. R. Wotton and C. Jeffs, *Biol. Conservat.*, 2001, **101**, 361–373.

147. N. W. Sotherton in "The Ecology of Temperate Cereal Fields", L. G. Firbank, N. Carter, J. F. Darbyshire and G. R. Potts (eds), Blackwell Scientific Publications, Oxford, 1991, 373–397.

148. C. Stoate, I. G. Henderson and D. M. B. Parish, *Ibis*, 2004, **146 (supplement 2)**, 203–209.
149. D. Smallshire, P. Robertson and P. Thompson, *Ibis*, 2004, **146 (supplement 2)**, 250–258.
150. D. Curry, "Farming and Food, a Sustainable Future", Report of the Policy Commission on the Future of Farming and Food, Defra, London, 2002.
151. Anon., "The environmental implications of the 2004 CAP reforms in England", Central Science Laboratory & Countryside and Community Research Unit, 2006, 62.

The Extent and Future of Global Insect Diversity

JESSICA J. HELLMANN AND NATHAN J. SANDERS

1 Introduction

The world is changing at an astonishing pace. Not since the emergence of life in the oceans has the Earth been modified so greatly and so quickly by one species – *Homo sapiens*.[1,2] Humans harvest a large fraction of global primary productivity;[3,4] we domesticate other species and grow them over vast areas in monoculture;[5] we release pollutants into the atmosphere that were stored underground millions of years ago;[6,7] and we knowingly and unknowingly spread exotic organisms around the globe.[8,9] Each of these changes has the potential to affect the global species pool, change local populations and cause extinctions. The impacts of anthropogenic change could be positive or negative for each affected species, but one thing is for certain – society ought to know what the consequences of such dramatic changes will be. If we do not understand the consequences of anthropogenic forces in the environment, we cannot take any steps to reduce or mitigate the negative effects of humanity on nature.

What do human-caused environmental changes mean for the species that inhabit this planet? To answer this question, we must know several essential pieces of information about biodiversity: where does it occur (*i.e.* what are the major biogeographic patterns on Earth?); how does it respond to global changes (*i.e.* what characteristics make particular taxa responsive or nonresponsive to changes in the environment?); and what types of change is it experiencing (*i.e.* what are the major threats that species face and how do those threats interact?). Clearly, providing even superficial answers to these questions is a daunting task, and entire careers of researchers are devoted to addressing only parts of these questions. Nevertheless, from time to time scientists must ask themselves where they stand on these big questions – how much have we learned and what do we still need to know? Such surveys of scientific knowledge

Issues in Environmental Science and Technology, No. 25
Biodiversity Under Threat
Edited by RE Hester and RM Harrison
© The Royal Society of Chemistry, 2007

can form the basis of future research agendas and can motivate renewed effort to address questions that have long evaded quantification. In this chapter, we attempt such a survey for the most diverse and most abundant eukaryotic organisms on the planet – the insects.

New insect species, even entire orders of insects, are being discovered all the time.[10] It is impossible to predict accurately how many remain to be discovered, and whole regions of the planet have been incompletely surveyed for insect diversity and ecology. Individual ecologists do not typically have the funds or personnel to perform some of the legwork that is necessary to discover new species, uncover global patterns and thoroughly catalogue ecological threats. Nor are there many multi-person, organised projects to do so worldwide.[11]

It is shocking and, in many ways, depressing that science has sent astronauts to the moon and has extended the human lifespan by decades, but we lack a simple understanding of the number and type of organisms with which we share this planet. As will become evident, we cannot solve that shortcoming in this review. Instead, we synthesise what is known about the diversity and function of several insect groups. In pointing out the gaps in our knowledge about the causes and consequences of insect diversity for the functioning of ecosystems, we provide a guide for future research. Specifically, we focus on terrestrial species and on population size and species extinction. This reduces our task but ignores the abundant insect fauna in aquatic ecosystems and the myriad genetic and other non-extinction effects caused by global change. These are, of course, no less important, but we leave those factors for others to consider.

2   A Diversity of Species and Functions

Insects are small-bodied arthropods that inhabit terrestrial and freshwater ecosystems. (They rarely occur in marine systems with the exception of some intertidal and coastal regions.) Flight is a key ancestral trait, enabling their colonisation and use of a wide range of habitats. As ectothermic animals, insects are generally sensitive to environmental conditions, thus serving as useful indicators for many forms of environmental change. Below we briefly examine the broad range of ecological characteristics in insects to demonstrate their dominance on Earth and their potential significance to life processes worldwide.

First, insects have a high reproductive output and short generation times relative to other animals. These traits typically confer large population sizes and relatively rapid evolution and adaptation. Parthenogenesis (asexual reproduction) also can play an important role in rapidly increasing population size, and it occurs in many species.[12] This population potential makes insects tractable in ecological research.

One of the most notable features of insects is their large surface area to volume ratio. This high ratio puts them at risk of desiccation, thereby driving patterns of increased species richness in moist environments.[12] Yet, insects inhabit a wide range of moisture gradients and temperatures from the dry heat of deserts (> 50°C) to dry, freezing temperatures below 30°C,[13,14] and many disperse widely among different environments. To withstand environmental extremes, insects either tolerate the conditions or avoid the precise times and locations where extremes occur, and they achieve this using morphological, physiological or behavioural adaptations. Such adaptations include morphologies for minimising convective heat loss or gain, supercooling at temperatures below freezing and living underground. Numerous species also enter periods of quiescence or diapause as a strategy to avoid unfavourable conditions. Long-distance dispersal sometimes is accomplished with the aid of wind or non-powered flight.

Key to understanding the astonishing adaptations of insects is an appreciation of their long evolutionary history.[15] This history also has allowed, in part, their divergence into varied ecological roles and lifestyles. For example, some insects scavenge on dead organic matter, some feed on green plants and some are predatory or parasitic. Their mouth-parts include structures adapted for chewing, piercing, gnawing and sucking.

Nearly half of all insects are phytophagous, or plant-eating, and insects in the six largest orders derive much or most of their food from plants. Phytophagy has evolved repeatedly in the insect clade, probably from scavenging.[12] From the plant's perspective, this relationship is either negative or positive – causing reduced plant biomass or enabling pollination and seed dispersal. The sophisticated interaction between insects and plants has played a role in the diversification of plants and the evolution of insect structures.[13] In contrast to phytophagy, insects with a predatory lifestyle appear to be less diverse. In a thorough study of the insects of the British Isles, for example, less than 4% of insect species were predatory.[16,17] More diverse than predatory but less diverse than herbivorous, approximately 35% of the British Isles insect fauna is parasitic on animals. In poorly sampled areas, it is possible that the proportion of parasitic species could be higher, but we know about the biology, diversity and natural history of only a few important parasitic taxa (*e.g.* mosquitoes). Both predators and parasites have participated in complex co-evolutionary associations with host species and pathogens.[18]

This diversity of ecological roles among the insects makes summarising their response to global change difficult, but a consistently high level of specialization,[11] small body size and ectothermic nature enable some generality. We focus on these broad-brush traits to explore the risk of global change to insects generally. Although some species may flourish as humans modify the environment (*e.g.* red imported fire ants, cockroaches, *etc.*), others probably will become extinct, and the loss of species diversity will have broad consequences for the function of ecosystems and the services they provide to humanity.

3 Services Provided by Insects

The potential consequences of insect species loss stem, in part, from the abundant ecosystem services that insects provide, including pollination, food for higher trophic levels, including humans, control of weeds and insect pests, soil improvement, decomposition, seed dispersal and beneficial gene sequences and/or genetically based products. Insects also have the power to defoliate large swaths of forest or cropland, thereby affecting ecosystem processes such as nutrient cycling and fire frequency.[19] Though some insects are pests or transmit pathogens, many are beneficial – either directly or indirectly. For example, insects contribute more to agricultural value than they remove or degrade,[12] and insects pollinate at least 177 crops worldwide.[20] In the USA, ants also disperse the seeds of >50% of plant species in eastern deciduous forests.[21] A few authors have attempted to quantify the economic value of such insect functions to human society. Calculations by Losey and Vaughan,[22] for example, suggest that native dung beetles contribute $57 billion to the US economy simply from the burial of cattle dung! They also estimate that native, wild insects contribute $3 billion in US pollination services annually,[23] $4.5 billion for pest control of native herbivores and $50 billion in support of recreational activities such as fishing and wildlife-watching. Though this view of the importance of insects is highly human-centric, this rationale alone argues for careful stewardship of insect diversity.

Most ecological services are provided by insects that are either dominant (abundant) or that play a unique ecological role. Rare or endangered insects probably contribute to ecosystem function in a minor way, but we typically know little about their ecological roles.[22] Further, the abundance and role of insects – including those that provide ecological services – vary in space and time. For example, an insect can be a pest in one portion of its range but regulated or even rare in another place and time.[24,25] There also is some preliminary indication that species providing important functions are particularly sensitive to changes in the environment, including the transformational changes to ecological systems that we discuss below.[26,27]

The complexity of life histories and range of functions in insects has led some to argue for a process-based perspective on insect conservation and an eye to preserving diverse habitats and species interactions.[28] This perspective argues for conservation at large spatial scales – areas large enough to support insect activities but small enough to apply management techniques and tools.[28] A fine-scale approach to conservation may be appropriate in the preservation of individual, endangered species but is unlikely to capture the range of conditions necessary to maintain insect diversity. With a functional perspective in mind, maximising diversity is probably a wise conservation goal. From an ethical and aesthetic perspective, it also is easy to argue that a springtail or a locust is just as important as charismatic megafauna such as rhinos or lions.[29]

4   Global Patterns of Insect Diversity

The sheer abundance and diversity of insects, the product of 400 million years of evolution, is amazing. For example, locust swarms can consist of millions of individuals per hectare, and a single ant colony can contain millions or billions of workers.[30] In one square metre of pastureland, over 100 000 springtails have been collected.[31]

Of all described species, insects comprise somewhere between 80 and 95% of diversity, approximately two million named species.[32] Obtaining an estimate of the actual number of insect species has proven difficult, however. Erwin's[33] initial estimates of global diversity based on real empirical data (and some whopping assumptions) elicited a flurry of other attempts and refutations.[34] Erwin suggested that there could be up to 30 million insect species, but others say that the number is probably closer to 10 million.[35,36] Regardless of whether the total number of insect species is 2 million or 100 million, it means that we still have many species to describe. The principal factor limiting accurate estimates of species diversity is a lack of basic information on the taxonomy, distribution and biology of insects. There are concerted efforts to catalogue insect diversity, but not necessarily to tackle the Herculean taxonomic tasks of describing and naming all species.[37] Our view is that describing new insect species and providing precise taxonomic certainty are no longer necessary or sufficient for the conservation of insect diversity. Instead, knowing where that diversity occurs and how it varies spatially seems to be a much more efficient use of limited resources.

So where do insects live? The short answer is: in every terrestrial habitat on Earth, from near the North Pole to the South Pole, from Death Valley to extreme elevations in the Himalayas; in caves, salt lakes and pools of petroleum; from tens of metres deep in the soil to the very tops of the world's largest trees. Insects are indeed everywhere, but they are not equally everywhere. That is, both richness and abundance vary spatially and temporally.

Like most taxa, insects are more diverse in the tropics than in temperate regions, insect diversity varies along elevational gradients and there are more insect species in larger areas than in smaller areas.[34,38,40] The most striking pattern, the latitudinal gradient, has intrigued biologists since the days of Darwin, Wallace and Bates. But only relatively recently have there been quantitative assessments of the latitudinal gradient in diversity.[41,42]

Though the latitudinal gradient has intrigued some of the best minds in ecology (there are ~2000 primary studies on diversity gradients in a variety of taxa), we still lack a fundamental understanding of why there are more species in the tropics than in temperate systems. Moreover, there are nearly as many hypotheses to explain the pattern as there are ecologists working on the pattern. Of the dozens of hypotheses to explain the latitudinal gradient in diversity, the following three are commonly cited. (1) It could be that the amount of available energy decreases with increasing latitude.[43] In the tropics, more energy, in the form of photosynthate, could lead to more individuals and more species. To most ecologists, this is likely to be seen as a leading cause of diversity gradients.[44] (2) Some authors have suggested that the latitudinal gradient arises because plant diversity is higher in the tropics. It has long been known that insect diversity is often correlated with plant diversity.[45] Thus, plant diversity may beget insect diversity by providing greater opportunities for specialisation. Another possibility is that the number of species per host plant is higher in the tropics.[36] However, global analyses of the number of insect species associated with plant species have found little evidence that the number of insects per plant could drive the latitudinal gradient.[46] Furthermore, disentangling the myriad ways in which plant diversity could drive insect diversity probably will require phylogenetically comparable plant communities in temperate and tropical sites.[11] (3) Some authors propose macroecological or metabolic explanations of the latitudinal gradient. These explanations posit that temperature might affect the speciation rate[47] or that body size–abundance relationships might create diversity gradients. These theories have received considerable theoretical, but little empirical, attention.[48,49]

Insect diversity also varies with elevation. Until recently, the general perception was that diversity declined monotonically with altitude.[39] But the patterns along elevational gradients vary among insect taxa, with nearly equal numbers of studies showing that diversity peaks at mid-elevations and that diversity declines with elevation.[50] Now that ecologists and biogeographers have at least begun to document the patterns in elevational diversity, the next critical step is to determine the underlying mechanisms that shape elevational diversity gradients. Many of the same mechanisms that shape latitudinal gradients in insect diversity probably also shape elevational gradients in diversity,[51,52] but that need not necessarily be the case. In fact, the mechanisms that shape diversity gradients, be they along elevational or latitudinal gradients, depend on the spatial and temporal scale that one considers.[50,53]

How close are we to knowing "the" mechanism that shapes insect diversity gradients? Our view is that as more data become available, from intensive surveys as well as climate data from increasingly sophisticated satellite imagery, our understanding of interacting mechanisms will be increased. There is no reason to think that a single factor drives all diversity gradients of the millions of insect species. Additionally, there are insect taxa for which the opposite trend – higher diversity in temperate zones than in the tropics – is observed (*e.g.* aphids, parasitic wasps).[16]

5   Threats to Insects Worldwide

If there are so many insect species, how can they possibly be threatened by human processes? And, of all of the organisms on Earth, why should we be concerned about the loss of the small-bodied and generally numerous insects? As we discuss above, insects provide essential functions in nature. As a group they are abundant, but as individual and specialised taxa, many of them are affected strongly by the environment and therefore may risk local or global extinction. This extinction risk arises from a number of factors including specialisation, small or widely fluctuating population sizes, sensitivity to thermal and chemical conditions in the environment and dependence on other species for basic life functions (see Table 1).

Table 1 Characteristics of insects likely to be high *vs.* low risk from global change, including land-use change, climate change and invasive species.

High risk	*Lower risk*
Small population size	Large population size
Narrow geographic range	Large geographic range
Widely fluctuating population size (exogenous population dynamics)	Regulated or stable population size (endogenous population dynamics)
Resource/habitat specialisation	Resource/habitat generalist
Narrow environmental tolerances (*e.g.* thermal tolerance)	Broad environmental tolerances and adaptive strategies for avoiding harsh conditions
High trophic position (*e.g.* parasitoid)	Basal tropic position (*e.g.* scavenger or plant-feeder)
Limited dispersal ability (*e.g.* wingless or small body size with limited flight distance)	High dispersal ability (*e.g.* winged with large flight muscles)
Involved in mutualism	Not dependent on mutualistic association with other organisms
Example: a small-bodied, specialised parasitoid with a small geographic distribution	Example: strong-flying butterfly that feeds on a number of abundant host plants and inhabits an entire continent

A number of studies indicate the role of early humans in influencing insect populations, through changes in the hydrological regime[54,55] and agricultural clearing,[28,56] for example. These changes were negative for some species, driving regional species losses, but were positive for other species. For example, agricultural clearing was largely considered beneficial for many species of Lepidoptera in Europe as they favour open conditions to closed forest in temperate regions.[57] As the various factors of global change interact and the extent of human modification of the environment grows, however, species losses may compound, further reducing global diversity.

Is there evidence of an emerging extinction crisis among the insect taxa? As with the debate over total insect diversity (above), there is little consensus about the rate of species loss within the insect clade.[29] Confirming a species loss also is dubious in insects because they are relatively hard to sample, many have small population sizes and many blink on-and-off in distinct habitat patches across a landscape.[28] Nonetheless, some authors suspect that 11 200 species have become extinct since 1600 AD and that perhaps at least 57 000 will become extinct in the next fifty years.[29,58] This rate of change, though uncertain, demands explanation and understanding so that future losses can be curbed or slowed by informed conservation management.

A variety of human-driven changes in the environment affect insects. Here we discuss three profound, and global, forms of environmental change: land use change, climate change and invasive species. In each case, we discuss potential vulnerabilities in insects and known responses.

6   Land-use Change

For those insects dependent on a particular habitat type, habitat loss due to land conversion means reduced availability of necessary resources. With reduced resources, we expect reduced population size and concordant species richness declines with incremental losses in habitat area. The principal drivers of habitat loss are agricultural conversion, land degradation from unsustainable agriculture, logging, urbanisation and human population growth. Habitat fragmentation, in contrast, is a by-product of habitat destruction. Some patches of habitat remain after a landscape has been modified, and these patches exist in a matrix of human-dominated ecosystems. When native patches are isolated from one another, each patch is vulnerable to penetration from organisms that persist in the matrix. Many of these organisms prey upon or compete with species that live in the patch.[59] With reductions in patch size and isolation from other patches, insects also move among patches less frequently.[60,61] For species that use a larger landscape to ensure regional persistence, this inability to disperse can cause decline or extinction.[62] We also expect isolated patches to have reduced species diversity relative to patches that are close together due to this process of dispersal limitation.

Using the traits in Table 1, we would expect species with specific habitat requirements and limited dispersal ability to be the most vulnerable to habitat loss and fragmentation.[63] The former implies harm from habitat degradation and an inability to persist in novel habitats; the latter implies an inability to escape affected areas by colonising other, suitable habitats. Some species, however, benefit from fragmentation by thriving in the matrix or in the secondary habitat occurring at the edge of the native-non-native border.[64] With greater intensification of land-use change over larger and larger areas, we would expect the negative effects of land use change to steadily reduce diversity, leaving behind a community composed of species with low-risk traits (Table 1).

Habitat loss and fragmentation take place in a wide variety of ecosystems, all of which contain insects. Many grassland and forest ecosystems, for example, have been converted to agriculture, forests have been lost to intensive logging and human settlement, and streams have been diverted or fragmented from hydrological projects. Habitat loss and fragmentation in the tropics, however, deserves special consideration as the diversity of insects occurring there is particularly large.[33,65] To make a rough, back-of-the-envelope calculation of species affected by tropical deforestation, we use a well-documented relationship in ecology called the species–area relationship. (This exercise was performed previously by Pimm[66] using estimates of deforestation that have since been updated.) The species–area relationship predicts the number of species occurring in a region using the surface area of that region and two empirically estimated parameters. For insect species living in continental areas, one of these two parameters – the one describing incremental losses in species richness with incremental losses in habitat area – falls in the range of 0.4 to 1.2, with an average of 0.7.[67] Plugging in the annual area lost to deforestation of 5.8 million ha year^{-1} in the 1990s[68] and using the average parameter value above, we predict extinction or severe declines in 4 million tropical forest species per year. Assuming that half of these are insects (probably a conservative value for the tropics) and that most tropical species are associated primarily with forest, we predict that 2 million insect species are impacted by tropical deforestation on an annual basis. (See also Ney-Nifle and Mangel[69] for a more sophisticated but similar approach to estimate species losses following deforestation.)

It is not clear if or how quickly these insect species will become extinct, however. Some may be doomed to extinction but hang on for some period of time due to high initial population sizes.[70,71] Others may persist elsewhere in the tropics if they have a large geographic range – that is, until a wider and wider region of tropical forest is logged and fragmented. At a constant rate of 5.8 million hectares deforested per year and a stock of 1150×10^6 hectares in 1990,[68] virgin or near-virgin tropical forest will be completely destroyed within 200 years and with it many of the insects species that reside there.

A number of studies support our back-of-the-envelope calculation that insect populations and species decline sharply in response to deforestation. Studies involving butterflies and termites, for example, indicate declines in species richness following the removal of tropical forest.[65,72,73,74,75] Still other studies suggest, however, that the quality of remaining patches is critical as is the dispersal distance of the individual affected species and historical land use in the area (*e.g.* Europe has long been converted to managed grassland, so the effects of modern habitat loss may be different or less than original loss).[76] The role of these mediating factors must be elucidated with further research, but a recent review of 20 experimental studies suggested that insects are consistent with theoretical expectations of land-use change.[77] The small body-size and short generation times of arthropods probably are the cause for the congruence of theory and empirical results.

As discussed above, a majority of insect species depends on plants for food resources, and we expect pronounced and direct changes in phytophagous species following land-use change. This expectation follows from the simple observation that the direct and immediate consequence of land use change is a shift in the plant community. Studies by Koricheva *et al.*,[78] for example, suggest that sessile species and specialists were strongly affected by changes in plant diversity in Europe, and plant species composition was the strongest factor in determining insect species change following land-use change. Such results suggest that much conservation of insect diversity can be accomplished by maximising the conservation of plant diversity.[28] This generalisation may even apply to species that indirectly consume plant material (*e.g.* scavengers) as leaf quality and the diversity of leaf species present in the leaf litter may maximise the diversity of the decomposing community.[79]

7 Climate Change

Increases in emissions of greenhouse gases from fossil fuel combustion and land-use conversion are changing regional patterns of temperature and precipitation.[6,7] In general, the climate is expected to become warmer and wetter, but the intensity of warming and precipitation will vary strongly by geographic location such that some areas actually will become dryer or cooler. Climate change is truly transformational – every organism will experience climate change – and many systems are responding to climatic modifications already.[80,81] The challenge for ecologists is understanding what features put populations and species at risk of extinction, what ecological processes might be disrupted and what conservation steps can be taken to mitigate negative impacts of climate change. As with land-use change above, potentially severe consequences of climate change argue for policies that slow or reverse carbon emissions, thereby preventing irreversible ecological damage. Using the traits in Table 1, we expect species with limited environmental tolerances (particularly thermal tolerances), resource specialists and species with specialised nutrient requirements to be the most vulnerable to climate change.[82–84] By vulnerable, we mean those most likely to experience population declines – declines that could lead to local extinctions, geographic range shifts or contractions, or even species losses.

The geographic ranges of several insect species have shifted in recent times, while others have contracted due to local population losses.[85,86] A range shift occurs when the equatorial edge of a geographic distribution becomes unsuitable while the poleward edge increases due to a shifting mean climate. That is, conditions systematically change across all of the habitats occupied by a species, pushing it to higher latitudes. Range contractions occur if populations decline at the equatorial and poleward range edges, and holes can appear within a species' range if centrally located populations are severely perturbed by climate change.

Changes in the distribution of functionally important species, endangered species, disease vectors and pest species all have potential ramifications for society. A recent study indicated that 13–85% of butterflies and other invertebrates could be threatened with extinction under climate change within 50 years, assuming a low-to-moderate amount of change and limited dispersal in the affected organisms.[87] This projection is based on the known geographic ranges of insects in Central America, South Africa and Australia; it estimates the distribution of these species pre- and post-climate change and uses the difference in occupied area in calculations with the species–area relationship. For all species on average, the authors predict that 22–37% of taxa will become extinct from climate change, suggesting that insects might be particularly vulnerable by comparison with other species. As with the rough calculation of extinction risk for land-use change above, these extrapolations are preliminary, but they are based on a well-documented empirical relationship. The sheer magnitude of these values suggests that much work needs to be done to understand the processes driving such dramatic change.

The view of climate change as a driving force of extinction, however, should be contrasted with its potential to cause increases in some species. Increases are likely to occur in species that tolerate changing conditions and exploit new niches as they come available. These species may have traits like those listed in the second column of Table 1. Unfortunately, some of the species that humanity fears the most – those with economic or human health consequences – might possess these very traits.[88] For example, evidence is accumulating that forest pest species are benefiting from climate change in regions where they did not previously occur and are causing extensive economic damage. The mountain pine beetle (*Dendroctonus ponderosae*) is a species that causes mortality of western pines in the USA. Potential hosts for this species extend into the Yukon and Northwest Territories, Canada, but outbreaks of the beetle have been restricted to southern British Columbia, Canada. Historically, winter temperatures in more northerly climates are thought to be too low for population persistence of the beetle, and extreme low temperatures are thought to end periods of outbreak in the historical beetle distribution.[89,90] Global warming may release this limitation on northward range expansion, however, enabling geographic spread of beetles and lengthening their outbreaks.[86] Recent records support this conclusion as pine beetle populations in British Columbia are steadily growing and damage has accumulated to record levels.[91] The outbreak probably will end with the depletion of suitable hosts.

Large-scale processes such as range shifts, range contractions, and altered pest–host relationships emerge because of localised processes playing out in many specific locations. A number of local factors can be important, but principle among them are the direct effects of changing climate on insects themselves. For example, the growth rate of individual insects varies linearly with temperature over a range of non-lethal values.[92] The slope and range of this linear function varies adaptively among species and potentially varies among populations within a species[93,94] The linear relationship leads to the expectation that warmer temperatures will increase individual performance and insect population size,[95,96] at least for temperatures up to the lethal limit. Extreme heat – those temperatures near the lethal maximum – also is predicted, either if the climate warms on average or if it becomes more variable.[6] High heat causes the denaturation of insect proteins, decreasing insect performance, and ultimately death.[97] Some insects have adaptive traits for physiologically moderating such conditions or behavioural mechanisms for avoiding them. To a degree, therefore, behavioural or developmental plasticity may reduce the direct effects of extreme temperature.

Temperature also has an effect on insect activity,[98] potentially affecting the ecological functions that insects perform. Individual insects are able to move about in the environment when temperature conditions fall within a narrow range.[99] Some species – including honeybees – extend the lower limit of this range with muscular shivering, but the body temperature of most insects is a simple function of ambient temperature. At the upper end of tolerable temperatures, for example, desert ants are active in the morning and evening so that the hottest hours of the day are spent in underground colonies.[100] Increased daytime temperature, therefore, could shift or modify available foraging times, affecting the dispersal of seeds. The activity of day-flying insects that pollinate plants also could be altered if temperatures shift outside the temperature range enabling flight. Figure 1, for example, shows the relationship

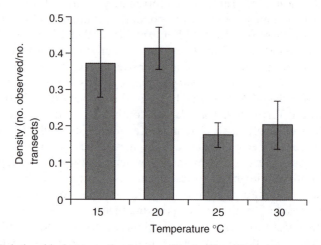

Figure 1   Relationship between the density of butterflies (*Eyrnnis propertius*) observed
in flight and the ambient temperature recorded near the ground.[101] Bars are
standard error. Data were recorded over three years at six locales on Van-
couver Island, British Columbia, near the edge of the species range. Researchers counted flying butterflies along survey transects during sunny
conditions. The number of 30 m transects surveyed in each site was standardised by habitat area so that density comparisons could be made among
sites. A non-parametric statistical test equivalent to ANOVA (Kruskal–
Wallis) indicates that butterfly density varies with temperature (p = 0.02)
such that fewer butterflies are active at high temperatures. Survey date also
describes some of the variation in butterfly density with fewer butterflies
observed later in the season as temperatures increase and the end of the
butterfly flight season approaches (regression analysis of density versus date:
$R^2 = 0.08$, n = 97, p = 0.005).

between ambient temperature and the density of butterflies (*Erynnis propertius*)
observed in flight during three years of surveys.[101] Observations were made
only when conditions were warm enough to sustain flight, but extremely high
temperatures correlated with diminished flight. If diminished flight occurs
across many species, pollination and other functions may decline, affecting
plant fitness and ecosystem productivity. If flight is diminished at the poleward
edge of a species range, this process could limit colonisation and range shifts in
flying insect species.

Reductions in insect population size also diminish activity-based ecosystem
functions. A simple survey of the density of native, seed-harvesting ants in
Eastern North America, for example, indicates declining population sizes at
low elevations, presumably in response to increasing temperatures (Figure 2). If
such species become locally extinct, the consequences, in terms of seed dispersal
and nutrient cycling, could cascade across trophic levels.

The indirect effects of climate change on insects could be as or more
pronounced than the direct effects. Indirect effects are those changes modified
by other species with which an insect interacts. For example, the timing of plant

Figure 2 Range shifts of the forest ant *Aphaenogaster rudis* based on MaxEnt models[133] incorporating 78 occurrence points and climate variables considered to be important to determine distributions of ants at large spatial scales, including: mean annual precipitation, mean annual temperature, maximum temperature of the warmest month, minimum temperature of the coldest month and precipitation of the wettest/driest months. Future conditions represent the predicted mean conditions from the Canadian Climate Center GCM (CCM3) in the year 2100 under an assumption of doubling of CO_2 concentrations downscaled to *c.* 1 km resolution from *c.* 50 km using spatial interpolation. Grey area shows the predicted current distribution, and black area shows the predicted future distribution in 2100.

development may change at a different rate under climate change than insect development. Assuming historical adaptation of insects to the timing of their host plants, particularly in temperate regions, asynchrony with plants could reduce the developmental success of some herbivores.[95] As half of the world's insect fauna consume plants, this type of indirect effect could profoundly alter insect diversity in seasonal environments.

Changes in the atmospheric concentration of greenhouse gases, notably carbon dioxide, also may indirectly affect insect herbivores. Increases in atmospheric carbon enhance plant growth by increasing the availability of an essential element in photosynthesis. Most studies have found changes in the carbon-to-nitrogen ratio in leaf tissue under elevated CO_2, suggesting that herbivores will need to eat more plant biomass to acquire a constant amount of nitrogen.[102] Limitations on available foraging time, therefore, could reduce insect performance.[103] Research is mixed on the effect of CO_2 to change the concentration of chemical defences and other plant compounds that affect plant quality.[104] Further study in this area is needed so that general results of CO_2 effects on plant-feeding insects can be predicted and managed.

8   Invasive Species

Though some authors suggest that the threats posed by invasive species may be small in comparison to other forms of global change,[105,106] invasive species are known to threaten terrestrial, freshwater and marine ecosystems worldwide.[8] Thirteen of the one hundred world's worst invasive species are insects; of those thirteen, five are ants.[107] Such a high number of invasive insects is potentially consequential because invasive insects can affect other species with which they compete or prey upon. For example, the invasive fire ant reduced native ant biodiversity by 70% and non-ant arthropods by 30% in Texas.[108] Similarly, the invasive Argentine ant reduced native ant diversity by at least 65% in California.[109]

Other well-known invasive insects cause declines in native diversity,[110] inflict economic damages in terms of crops and forest products[111] and demand economic resources by households and governments to mitigate their effects or attempt their control. For example, billions of dollars have been spent in the USA to control the red imported fire ant, to no avail.[111] The prominent conservationist and myrmecologist E. O. Wilson has quipped that the failed attempts to control the fire ant are "the Vietnam of entomology".[112] A study using a well-known empirical relationship in ecology called the species accumulation curve, together with the relationship between the number of introduced insects and the amount of US imports, calculated that the number of species introduced to the US will grow over time.[113] Fitting historical data on invasive insects, this study suggests that at least 115 new insect species are likely to be introduced to the USA by 2020.

Reducing the flow of insect invaders will be difficult as their small size, ability to persist in a range of materials and diapause strategies facilitate their transport by people and industry. Regulatory agencies in the USA, for example, such as USDA APHIS are charged with protecting US agriculture from insect invaders, but their methods of permitting and sporadic inspection are probably insufficient given the potential pathways of introduction and the inconspicuous nature of many insects. Other voluntary, international agreements such as the International Plant Protection Convention impose policies such as fumigation treatment of wood pallets used in international shipping (Internal Standard for Phytosanitary Measure #15). These treatments aim to reduce the international transport of bark beetles with the potential to harm logging industries. The rate of bark beetle introduction by this method is estimated at one species year^{-1}, and a total of 25 bark beetle species have been introduced within the last 21 years.[114] Unlike the bark beetle, potential invaders with little or no economic value are generally unregulated.[115]

Invasive species that affect the structure of plant communities probably have the largest effect on native insect species by fundamentally changing the habitat itself, and such a transformation already is under way as 23% of plant species in the USA are non-native.[111] Local extinction of native plants used by specialist insect species also is a potential consequence of plant invasion.[116]

The invasions of insects and plants often are linked. For example, many insects were introduced as biological control agents to limit the spread or impact of invasive plant species. These attempts have met with limited success,[117] and some have even had detrimental non-target effects on native plant species.[118]

9 Where Do We Go from Here?

Insect diversity is everywhere, and much of humanity relies on insects to perform critical ecosystem functions. Insects are vulnerable to extinction, particularly in response to the transformational changes of habitat loss, climate change and invasive species. What is the best, most efficient path forward to document and understand the causes and consequences of insect biodiversity and its disappearance? We suggest that the following steps are needed to better understand insect diversity and its endangerment. We acknowledge that meeting these goals may be difficult, but science has conquered many difficult goals before. It's time to set high goals and achieve them in ecology also.

9.1 A New Taxonomy

Organismal biology is slowly losing taxonomic specialists who can perform identifications, and several authors have called for a new taxonomy specifically directed toward the study of risk assessment.[119-121] This new taxonomy would be more strategic – focusing on hyper-diverse or little explored areas. Moreover, rather than cataloguing new species in expensive and little-read journals, taxonomy and systematic surveys should be made freely available online to the conservation and ecological community.[122] This new taxonomy also should attempt to generalise from some groups to others – to derive general principles[123] – and it should provide morphological keys that are accessible to untrained practitioners on the ground. New technologies such as those involving portable, digital databases that can be taken into the field or devices that perform genetic-based identification on the fly are examples of new tools that someone performing an insect diversity survey might implement.

9.2 Systematic Sampling

Intensive sampling and local inventories of little-studied regions of the world, including sub-Saharan Africa, much of tropical Indonesia and the extreme forest canopies of old growth forests is much needed. Most ecologists and taxonomists live in developed countries and study patterns of diversity at relatively high latitudes. This new systematic sampling also could employ local peoples and establish a network of parataxonomists throughout the world. Technology transfer or funds for international collaboration between developed and developing countries is desperately needed to quantify diversity in hotspots of biodiversity. Quantitative tools for measuring and comparing diversity only now are emerging from their infancy and need further development[124] so that we can assess the extent to which particular locations have been undersampled and can more accurately compare diversity between and among locations.

9.3 Synthesis of Biodiversity Inventories

Because of limited funds, inherent interests and a fondness for particular locations, ecologists and entomologists often focus their research efforts on a particular taxon or on a particular region. The work gets published in peer-reviewed journals, but such data are infrequently combined and synthesised across locations. Such a combined database could be useful in initially assessing global patterns of biodiversity (at least until better surveys are available as described above).[121] Such a synthesis has been done for birds[125] – an interesting and well-studied group but one that is not as diverse as insects. We propose accomplishing the same task with particular insect taxa that are relatively well known and well studied. For example, a global database on butterfly diversity, compiled from field guides, ecological studies and collecting trips would be invaluable. To our knowledge, no such database exists except where compiled for individual regions of interest. Similarly, detailed faunistic surveys of particular places, such as the All Taxa Biodiversity Inventory in the Great Smoky Mountains National Park, USA, could be initiated at other locations.

9.4 Multi-factor Research

Because the various forms of global change act in concert, it will not be sufficient to measure and investigate each effect in isolation. A growing body of research shows that each of these factors can affect biodiversity in a synergistic way that may outweigh their independent effects.[126] Few experiments, however, manipulate multiple factors of global change in a factorial design. Such studies are needed to identify non-linearities and thresholds in species' responses. For example, studies that investigate multiple, interacting effects such as carbon and nitrogen fertilisation have been initiated, including studies that assess their impacts on herbivorous insects.[127–129] These studies also could be further integrated, for example, with the effect of resource availability due to habitat loss or invasive species.

9.5 Generating a Trait-based Understanding of Global Change

A particularly productive avenue for future research would be to identify the ecological traits – rather than the individual species – that make insects susceptible to habitat loss, climate change and invasive species. For example, species may vary in their ability to shift their geographic ranges under climate change in a way that can be predicted by their ecological traits and evolutionary history.[94] Invasion biologists also have been searching for traits associated with invasive species[130] and, to a lesser degree, susceptibility to invasive competitors and predators.[113] Complicating and confounding factors will mask the relationship between traits and functions, and these factors also must be understood. For example, particular ecological traits have been associated with some invasive species, but other taxa with these same characteristics have failed to take hold when introduced. Though potentially confounded with other factors and conditions, such trait-based studies offer the best opportunity for generalising among species and crafting management tools. Studies on the process of invasion itself, including interdisciplinary research on the pathways of human spread and the changes that globalisation and climate change will bring to these pathways, also are much needed. If general principles apply here, the flow of invasives might be slowed regardless of the species – and their traits – that compose that flow.

10 Conclusions

In this review, we have focused on transformative and global processes of environmental modification by humans. We have not considered factors such as local pollution (*e.g.* pesticide application) or natural disturbances such as fires. This is not to imply that these factors are not potentially important potential drivers of insect population losses and local declines in the functions and services that insects provide. Aquatic insects, for example, are known to be strongly affected by water pollution in regions near intensive agriculture and urbanisation.[131,132] Local factors, however, potentially have local solutions, and the negative effects of pollution and other anthropogenic disturbance agents could be mediated with local conservation attention.

The economic and demographic changes caused by global environmental change, in contrast, are vexing on the largest of scales. Society must ask itself: What kind of world do we want? What risks are we willing to impose on natural systems? Are we willing to accept the consequences of biomanipulation where some species benefit while others decline? Presently, *c.*10% of the terrestrial world is set aside for conservation. Does that 10% capture the diversity and ecosystem function of insects that need to be conserved? Until we address shortcomings in our understanding of insect species diversity and the consequences of extinctions, we cannot appreciate the full scope of global change. Until we re-evaluate and re-invent our society, we must live with the biological consequences – whether they are fully understood or not.

Acknowledgements

We thank Windy Bunn, Greg Crutsinger, Rob Dunn, Matt Fitzpatrick, Travis Marsico, Jillian Mueller, Maggie Patrick, Shannon Pelini and Kirsten Prior for thoughtful comments on a draft of this manuscript.

References

1. F. S. Chapin, E. S. Zavaleta, V. T. Eviner, R. L. Naylor, P. M. Vitousek, H. L. Reynold, D. U. Hooper, S. Lavorel, O. E. Sala, S. E. Hobbie, M. C. Mack and S. Díaz, *Nature*, 2000, **405**, 234.
2. W. V. Reid, H. A. Mooney, A. Cropper, D. Capistrano, S. R. Carpenter, K. Chopra, P. Dasgupta, T. Dietz, A. K. Duriappah, R. Hassan, R. Kasperson, R. Leemans, R. M. May, A. J. McMichael, P. Pingali, C. Samper, R. Scholes, R. T. Watson, A. H. Zakri, Z. Shidoing, N. J. Ash, E. Bennett, P. Kumar, M. J. Lee, C. Raudepp-Hearne, H. Simons, J. Thonell and M. B. Zurek (eds), "Ecosystems and Human Well-being: Synthesis", Island Press, Washington DC, 2005.
3. P. M. Vitousek, P. R. Ehrlich, A. H. Ehrlich and P. A. Matson, *Bioscience*, 1986, **34**, 368.
4. M. L. Imhoff, L. Bounoua, T. H. Rickets, C. Loucks, R. Harris and W. T. Lawrence, *Nature*, 2004, **429**, 870.
5. D. Tilman, R. M. May, C. L. Lehman and M. A. Nowak, *Nature*, 1994, **371**, 65.
6. J. T. Houghton and Y. Ding, "Climate Change 2001: the Scientific Basis. Contribution of Working Group I to the Third Assessment Report of the Intergovernmental Panel on Climate Change", Cambridge University Press, Cambridge, UK, 2002.
7. J. T. Houghton, in "Global Environmental Change", R. E. Hester and R. M. Harrison (eds), *Issues Environ. Sci. Tech.*, 2002, **17**, 1.
8. R. N. Mack, D. Simberloff, W. M. Lonsdale, H. Evans, M. Clout and F. Bazzaz, *Ecol. Appl.*, 2000, **10**, 689.
9. P. E. Hulme, *this volume, Ch. 3*.
10. K. D. Klass, O. Zompro, N. P. Kristensen and J. Adis, *Science*, 2002, **296**, 1456.
11. V. Novotny, P. Drozd, S. E. Miller, M. Kulfan, M. Janda, Y. Basset and G. D. Weilblen, *Science*, 2006, **313**, 1115.
12. H. W. Daly, J. T. Doyen and A. H. Purcell III, "Introduction to Insect Biology and Diversity", Oxford University Press, Oxford, UK, 1998.
13. D. L. Denlinger and G. D. Yocum in "Temperature Sensitivity in Insects and Application in Integrated Pest Management", G. J. Hallman and D. L. Denlinger (Eds), Westview Press, Boulder, CO, 1998, 7.
14. B. J. Sinclair, P. Vernon, C. J. Klok and S. L. Chown, *Trends Ecol. Evol.*, 2003, **18**, 257.
15. P. R. Ehrlich and P. H. Raven, *Evolution*, 1964, **18**, 586.
16. P. W. Price, "Insect Ecology", John Wiley & Sons, New York, NY, 1997.
17. G. S. Kloet and W. D. Hincks, "A Checklist of British Insects", Kloet and Hincks, Stockport, UK, 1945.
18. R. M. Anderson and R. M. May, *Parasitology*, 1982, **85**, 411.
19. A. T. Classen, J. DeMarco, S. C. Hart, T. G. Whitmham, N. S. Bocc and G. W. Koch, *Soil Biol. Biochem.*, 2006, **38**, 972.

20. E. Crane, "Bees and Beekeeping: Science, Practice, and World Resources", Comstock Publishing Associates, Ithaca, NY, 1990.
21. A. J. Beattie, "The Evolutionary Ecology of Ant-Plant Mutualisms", Cambridge University Press, Cambridge, UK, 1985.
22. J. E. Losey and M. Vaughan, *Bioscience*, 2006, **56**, 311.
23. T. H. Ricketts, G. C. Daily, P. R. Ehrlich and C. D. Michener, *Proc. Nat. Acad. Sci. USA*, 2004, **101**, 12579.
24. B. E. Tabashnik, *Evolution*, 1983, **37**, 150.
25. C. Hambler and M. R. Speight, *Conservat. Biol.*, 1996, **10**, 892.
26. C. Kremen, N. M. Williams and R. W. Thorp, *Proc. Nat. Acad. Sci. USA*, 2002, **99**, 16812.
27. T. H. Larsen, N. M. Williams and C. Kremen, *Ecol. Lett.*, 2005, **8**, 538.
28. M. J. Samways, "Insect Diversity and Conservation", Cambridge University Press, Cambridge, UK, 2005.
29. R. R. Dunn, *Conservat. Biol.*, 2005, **19**, 1030.
30. T. Giraud, J. S. Pedersen and L. Keller, *Proc. Nat. Acad. Sci. USA*, 2002, **99**, 6075.
31. G. Salt, F. S. J. Hollick, F. Raw and M. V. Brian, *J. Anim. Ecol.*, 1948, **17**, 139.
32. B. Groombridge (ed.), "Global Diversity-Status of the Earth's Living Resources. Compiled by the World Conservation Monitoring Centre", Chapman and Hall, London, UK, 1992.
33. T. L. Erwin, *The Coleopterists Bulletin*, 1982, **36**, 47.
34. R. M. May, *Sci. Am.*, 1992, **267**, 42.
35. K. J. Gaston, *Conservat. Biol.*, 1991, **5**, 283.
36. F. Odegaard, *Biol. J. Linn. Soc.*, 2000, **71**, 583.
37. I. Oliver and A. J. Beattie, *Conservat. Biol.*, 1996, **10**, 99.
38. M. L. Rosenzweig, "Species Diversity in Space and Time", Cambridge University Press, Cambridge, UK, 1995.
39. C. Rahbek, *Ecography*, 1995, **18**, 200.
40. M. Kaspari, M. Yuan and L. E. Alonso, *Am. Nat.*, 2003, **161**, 459.
41. A. F. G. Dixon, P. Kindlmann, J. Leps and J. Holman, *Am. Nat.*, 1987, **129**, 580.
42. J. Kouki, P. Niemelä and M. Viitasaari, *Annales Zoologici Fennici*, 1994, **31**, 83.
43. D. J. Currie, G. G. Mittelbach, H. V. Cornell, R. Field, J. -F. Guégan, B. A. Hawkins, D. M. Kaufman, J. T. Kerr, T. Oberdorff, E. O'Brien and J. R. G. Turner, *Ecol. Lett.*, 2004, **7**, 1121.
44. K. L. Evans, P. H. Warren and K. J. Gaston, *Biol. Rev.*, 2005, **80**, 1.
45. T. R. E. Southwood, V. K. Brown and P. M. Reader, *Biol. J. Linn. Soc.*, 1979, **12**, 327.
46. K. J. Gaston, *Conservat. Biol.*, 1991, **5**, 283.
47. J. H. Brown, J. F. Gillooly, A. P. Allen, V. M. Savage and G. B. West, *Ecology*, 2004, **85**, 1771.
48. R. M. May, *Phil. Trans. Roy. Soc. London Series B-Biological Sciences*, 1990, **330**, 293.

49. E. Siemann, D. Tilman and J. Haarstad, *Nature*, 1996, **380**, 704.
50. C. Rahbek, *Ecol. Lett.*, 2005, **8**, 224.
51. N. J. Sanders, *Ecography*, 2002, **25**, 25.
52. N. J. Sanders, J. Mos and D. Wagner, *Global Ecol. Biogeogr.*, 2003, **12**, 93.
53. R. R. Dunn, C. M. McCain and N. J. Sanders, *Global Ecol. Biogeogr.*, in press.
54. P. Ponel, J. -L. de Beaulieu and K. Tobolski, *The Holocene*, 1992, **2**, 117.
55. V. Andrieu-Ponel and P. Ponel, *Biodiversity and Conservation*, 1999, **8**, 391.
56. P. J. Osbourne in "Studies in Quaternary Entomology-an Inordinate Fondness for Insects", A. C. Ashworth, P. C. Buckland and J. P. Sadler (eds), John Wiley and Sons, New York, NY, 1997, 193.
57. J. E. Thomas in "The Scientific Management of Temperate Communities for Conservation", I. F. Spellerber, F. B. Golsmith and M. G. Morris (eds), Oxford University Press, Oxford, UK, 1991, 149.
58. N. A. Mawdsley and N. E. Stork in "Insects in a Changing Environment", R. Harrington and N. E. Stork (eds), Academic Press, London, UK, 1995, 321.
59. A. D. Chalfoun, F. R. Thompson and M. J. Ratnaswamy, *Conservat. Biol.*, 2002, **16**, 306.
60. T. H. Ricketts, *Am. Nat.*, 2001, **158**, 87.
61. E. I. Damschen, N. M. Haddad, J. L. Orrock, J. J. Tewksbury and D. J. Levey, *Science*, 2006, **313**, 1284.
62. C. D. Thomas, *Proc. Roy. Soc. London Series B-Biological Sciences*, 2000, **267**, 139.
63. T. Tscharntke, I. Steffan-Dewenter, A. Kruess and C. Thies, *Ecol. Res.*, 2002, **17**, 229.
64. P. Duelli, P. M. Studer, I. Marchand and S. Jakob, *Biol. Conservat.*, 1990, **54**, 193.
65. A. D. Watt, N. E. Stork, P. Eggleton and D. S. Srivastava in "Forests and Insects", A. D. Watt, N. E. Stork and M. D. Hunter (eds), Chapman and Hall, London, UK, 1997, 273.
66. S. L. Pimm in "Conservation Science and Action", W. J. Sutherland (ed), Blackwell Science, Oxford, UK, 1998, 20.
67. M. D. Collins, D. P. Vazquez and N. J. Sanders, *Evol. Ecol. Res.*, 2002, **4**, 457.
68. F. Achard, H. D. Eva, H. -J. Stibig, P. Mayau, J. Gallego, T. Richards and J. -P. Malingreau, *Science*, 2002, **297**, 999.
69. M. Ney-Nifle and M. Mangel, *Conservat. Biol.*, 2000, **14**, 893.
70. C. Loehle and B. L. Li, *Ecol. Appl.*, 1996, **6**, 784.
71. D. Tilman, *Ecology*, 1999, **80**, 1455.
72. J. D. Holloway, A. H. Kirkspriggs and C. V. Khen, *Phil. Trans. Roy. Soc. London Series B-Biological Sciences*, 1992, **335**, 425.
73. J. K. Hill, K. C. Hamer, L. A. Lace and W. M. T. Banham, *J. Appl. Ecol.*, 1995, **32**, 754.

74. D. T. Jones, F. X. Susilo, D. E. Bignell, S. Hardiwinoto, A. N. Gullison and P. Eggleton, *J. Appl. Ecol.*, 2003, **40**, 380.
75. H. L. Vasconcelos, J. M. S. Vilhena, W. E. Magnusson and A. L. K. M. Albernaz, *J. Biogeogr.*, 2006, **33**, 1348.
76. J. Dauber, J. Bengtsson and L. Lenoir, *Conservat. Biol.*, 2006, **20**, 1150.
77. D. M. Debinski and R. D. Holt, *Conservat. Biol.*, 2000, **14**, 342.
78. J. Koricheva, P. H. Mulder, B. Schmid, J. Joshi and K. Huss-Danell, *Oecologia*, 2000, **125**, 271.
79. I. Armbrecht, I. Perfecto and J. Vandermeer, *Science*, 2004, **304**, 284.
80. T. L. Root, J. T. Price, K. R. Hall, S. H. Schneider, C. Rosenzweig and J. A. Pounds, *Nature*, 2003, **421**, 57.
81. C. Parmesan and G. Yohe, *Nature*, 2003, **421**, 37.
82. J. J. Hellmann in "Wildlife Responses to Climate Change: North American Case Studies", S. H. Schneider and T. L. Root (eds), Island Press, Washington DC, 2001, 93.
83. C. D. Thomas, E. J. Bodsworth, R. J. Wilson, A. D. Simmons, Z. G. Davies, M. Musche and L. Conradt, *Nature*, 2001, **411**, 577.
84. N. R. Andrew and L. Hughes, *Ecol. Entomol.*, 2004, **29**, 527.
85. C. Parmesan, N. Ryrholm, C. Stefanescu, J. K. Hill, C. D. Thomas, H. Descimon, B. Huntley, L. Kaila, J. Kullberg, T. Tammaru, W. J. Tennent, J. A. Thomas and M. Warren, *Nature*, 1999, **399**, 579.
86. J. F. McLaughlin, J. J. Hellmann, C. L. Boggs and P. R. Ehrlich, *Proc. Nat. Acad. Sci. USA*, 2002, **99**, 6070.
87. C. D. Thomas, A. Cameron, R. E. Green, M. Bakkenes, L. J. Beaumont, Y. C. Collingham, B. F. N. Erasmus, M. F. de Siqueira, A. Grainger, L. Hannah, L. Hughes, B. Huntley, A. S. van Jaarsveld, G. F. Midgley, L. Miles, M. A. Ortega-Huerta, A. T. Peterson, O. L. Philips and S. E. Williams, *Nature*, 2004, **427**, 145.
88. J. A. Logan, J. Regniere and J. A. Powell, *Front. Ecol. Environ.*, 2003, **1**, 130.
89. A. L. Carroll, S. W. Taylor, J. Régnière and L. Safranyik in "Mountain Pine Beetle Symposium: Challenges and Solutions", T. L. Shore, J. E. Brooks and J. E. Stone (eds), Pacific Forestry Center, Victoria, BC, Canada, 2003, 223.
90. M. J. Ungerer, M. P. Ayres and M. J. Lombardero, *J. Biogeogr.*, 1999, **26**, 1133.
91. M. Eng, A. Fall, J. Hughes, T. Shore, B. Riel, P. Hall and A. Walton, Canadian Forest Service and BC Forest Service, 2005.
92. N. Gilbert and D. A. Raworth, *Canadian Entomologist*, 1996, **128**, 1.
93. M. P. Ayres and J. M. Scriber, *Ecol. Monogr.*, 1994, **64**, 465.
94. J. J. Hellmann, S. M. Pelini and K. M. Prior, in review.
95. J. S. Bale, G. J. Masters, I. D. Hodkinson, C. Awmack, T. M. Bezemer, V. K. Brown, J. Butterflied, A. Buse, J. C. Coulson, J. Farrar, J. E. G. Good, R. Harrington, S. Hartley, T. F. Jones, R. L. Lindroth, M. C. Press, I. Symmioudis, A. D. Watt and J. B. Whittaker, *Global Change Biol.*, 2002, **8**, 1.

96. L. Crozier and G. Dwyer, *Am. Nat.*, 2006, **167**, 853.
97. G. N. Somero, *Ann. Rev. Physiol.*, 1995, **57**, 43.
98. B. Heinrich and P. H. Raven, *Science*, 1972, **176**, 597.
99. J. G. Kingsolver, *Physiol. Zool.*, 1989, **62**, 314.
100. B. Hölldobler and E. O. Wilson, "The Ants", The Belknap Press of Harvard University Press, Cambridge, MA, 1990.
101. K. M. Prior and J. J. Hellmann, in review.
102. M. C. Hall, P. Stiling, D. C. Moon, B. G. Drake and M. D. Hunter, *J. Chem. Ecol.*, 2005, **31**, 267.
103. P. Lundberg and M. Astrom, *Am. Nat.*, 1990, **135**, 547.
104. E. L. Zvereva and M. V. Kozlov, *Global Change Biol.*, 2006, **12**, 27.
105. M. Sagoff, *J. Agr. Environ. Ethics*, 2004, **18**, 215.
106. J. Gurevitch and D. K. Padilla, *Trends Ecol. Evol.*, 2004, **19**, 470.
107. Anon., *Global Invasive Species Database*, 2005.
108. S. D. Porter and D. A. Savignano, *Ecology*, 1990, **71**, 2095.
109. D. A. Holway, L. Lach, A. V. Suarez, N. D. Tsutsui and T. J. Case, *Ann. Rev. Ecol. Systemat.*, 2002, **33**, 181.
110. K. Schmidt, *New Scientist*, 1995, **148**, 28.
111. D. Pimentel, L. Lach, R. Zuniga and D. Morrison, *Bioscience*, 2000, **50**, 53.
112. J. Blu Buhs, "The Fire Ant Wars", University of Chicago Press, Chicago, IL, 2004.
113. J. M. Levine and C. M. D'Antonio, *Oikos*, 1999, **87**, 15.
114. R. A. Haack, *Can. J. Forest Res.*, 2006, **36**, 269.
115. D. M. Lodge, S. L. Williams, H. MacIsaac, K. Hayes, B. Leung, S. Reichard, R. N. Mack, P. B. Moyle, M. Smith, D. A. Andow, J. T. Carolton and A. McMichael, Position paper of the Ecological Society of America, Washington DC, USA, 2006.
116. M. J. Samways, P. M. Caldwell and R. Osborn, *Agr. Ecosyst. Environ.*, 1996, **59**, 19.
117. D. Simberloff and P. Stiling, *Ecology*, 1996, **77**, 1965.
118. S. M. Louda, D. Kendall, J. Connor and D. Simberloff, *Science*, 1997, **277**, 1088.
119. P. R. Ehrlich, "A World of Wounds: Ecologists and the Human Dilemma", Ecology Institute, Luhe, Germany, 1997.
120. M. R. Wilson, *Nature*, 2000, **407**, 559.
121. H. C. J. Godfray, *Nature*, 2002, **417**, 17.
122. N. J. Gotelli, *Phil. Trans. Roy. Soc. London Series B*, 2004, **359**, 585.
123. T. R. New in "The Other 99%: the Conservation and Biodiversity of Invertebrates", W. Ponder and D. Lunnedy (eds), Royal Zoological Society of New South Wales, Mosman, Australia, 1999, 154.
124. A. E. Magurran, "Measuring Biological Diversity", Blackwell Sciences, Malden, MA, 2004.
125. M. Pautasso and K. J. Gaston, *Ecol. Lett.*, 2005, **8**, 282.
126. J. F. Weltzin, R. T. Belote and N. J. Sanders, *Front. Ecol. Environ.*, 2003, **1**, 146.

127. J. E. Kerslake, S. J. Woodin and S. E. Hartley, *New Phytologist*, 1998, **140**, 43.

128. K. E. Percy, C. S. Awmakc, R. L. Lindroth, M. E. Kubiske, B. J. Kopper, J. G. Isebrands, K. S. Pregitzer, R. Hendre, R. E. Dickson, D. R. Zak, E. Oksanen, J. Sober, R. Harrington and D. F. Karnosky, *Nature*, 2002, **420**, 403.

129. E. A. Sudderth, K. A. Stinson and F. A. Bazzaz, *Global Change Biol.*, 2005, **11**, 1997.

130. C. S. Kolar and D. M. Lodge, *Trends Ecol. Evol.*, 2001, **16**, 199.

131. J. D. Allan and A. S. Flecker, *Bioscience*, 1993, **43**, 32.

132. D. A. Polhemus, *Am. Zool.*, 1993, **33**, 58.

133. S. J. Phillips, R. P. Anderson and R. E. Schapire, *Ecol. Model.*, 2006, **190**, 231.

Biological Invasions in Europe: Drivers, Pressures, States, Impacts and Responses

PHILIP E. HULME

1 Biological Invasions in Europe: a Framework for Best Practice

Biological invasions by non-native, "exotic" or "alien" species (see Table 1 for definitions) are large-scale phenomena of widespread importance and represent one of the major current threats to European biodiversity.[1] The introduction and transfer of alien invasive species among continents, regions and nations has often had significant impacts on the recipient aquatic and terrestrial ecosystems.[2] Yet a frequent perception is that, compared to other continents, the semi-natural ecosystems of Europe are more resistant to invasion as a result of the long interaction between humans and their environment[3] and the fact that intentional species introductions were undertaken far more frequently by European settlers colonising other continents than on their return to Europe.[4] Such a perception is in need of revision. The Industrial Revolution heralded an order of magnitude change in the scale of human impacts on the environment and globalisation of trade has increased the rate at which species are introduced into Europe. As a result, numerous non-native species, many introduced into Europe little more than 200 years ago, have become successfully established over large areas of the continent.

Europe's size, number of countries and free trade arrangements make it essential to promote consistency in approach against the threat posed by biological invasions and thus avoid unilateral national efforts being undermined by their neighbours' inaction. The "European Strategy on Invasive Alien Species" aims to promote the development and implementation of coordinated measures and efforts throughout the region to minimise the adverse impact of invasive alien species on Europe's biodiversity, economy and human health.[5] In support of the European Strategy, this review adopts an international standard framework for comparative assessments of threats to the environment. The

Issues in Environmental Science and Technology, No. 25
Biodiversity Under Threat
Edited by RE Hester and RM Harrison
© The Royal Society of Chemistry, 2007

DPSIR framework (Drivers-Pressures-States-Impacts-Responses) is widely used to assess and manage environmental problems, but has yet to be applied to biological invasions. The framework can be adapted for the specific threat of alien species on biodiversity where:

1. Drivers are the socio-economic and socio-cultural forces underpinning human *activities* that determine the *magnitude* of biological invasions
2. Pressures reflect the *exposure* of ecosystems to the threat of alien species
3. States are measurements of the *condition* of the environment in terms of the distribution and abundance of alien species
4. Impacts are the *effects* of alien species on biodiversity and ecosystem function

Table 1   Terms used to describe biological invasions.[11,34]

Term	Description
"introduction"	The movement, by human agency, of a species, subspecies or lower taxon (including any part, gametes or propagule that might survive and subsequently reproduce) outside its natural range (past or present). This movement can be either within a country or between countries.
"intentional introduction"	An introduction made deliberately by humans, involving the purposeful movement of a species outside of its natural range and dispersal potential. (Such introductions may be authorised or unauthorised.)
"unintentional introduction"	An unintended introduction made as a result of a species utilising humans or human delivery systems as vectors for dispersal outside its natural range.
"alien species"	A species, subspecies or lower taxon introduced outside of its natural range (past or present) and dispersal potential (*i.e.* outside the range it occupies naturally or could not occupy without direct or indirect introduction or care by humans) and includes any part, gametes or propagule of such species that might survive and subsequently reproduce. Synonyms include non-native, non-indigenous and exotic species.
"acclimatised species"	An alien species which becomes established in natural or semi-natural ecosystems but is incapable of maintaining free-living populations without human intervention.
"naturalised species"	An alien species which becomes established in natural or semi-natural ecosystems with free-living, self-maintaining and self-perpetuating populations unsupported by and independent of humans.
"feral species"	A naturalised alien species that has reverted to the wild from domesticated stock (*e.g.* has undergone some change in phenotype, genotype and/or behaviour as a result of artificial selection in captivity).
"invasive alien species"	A naturalised alien species which is an agent of change, and threatens human health, economy and/or native biological diversity.

5. Responses refers to the regulatory and strategic *actions* available to society to mitigate the threat of invasions in each of the four preceding framework components

Each of the framework components is dealt with in the following sections in order to present trends in a structured manner and thus facilitate communication to policy makers and the public. The results highlight several components where science and policy currently fail to address the threat of alien species to European biodiversity.

2 The Trouble with Trade and Travel: Economic Drivers of Biological Invasions

Europe is a major market for the import and export of international trade, and this commerce has facilitated the spread of alien species into and within the region through a diversity of means. These include intentional releases into the wild, accidental (but often foreseeable) escapes from captivity and/or domestication, as well as unintentional introductions as a by-product of trade. The use of alien species in farming, forestry, aquaculture and for recreational purposes has increased in much of Europe since the beginning of the 20th century. Alien species may be imported because they offer increased economic returns (*e.g.* plantations of fast-growing non-native conifers), satisfy demand for exotic products (*e.g.* fur trade), feed on and suppress other species (*e.g.* biological control agents) or simply because people like them (*e.g.* pets and many garden plants). Although there is considerable variation among taxa and biomes in the means by which alien taxa have entered Europe, many aliens have been introduced intentionally (Figure 1). Vertebrates have often been intentionally introduced into the wild as game animals (*e.g.* pheasant (*Phasianus colchicus*), fallow deer (*Dama dama*) and rainbow trout (*Oncorhynchus mykiss*)). Associated with intentional establishment of game are alien plants introduced for cover (*e.g.* snowberry (*Symphoricarpus albus*)) and invertebrates as bait (*e.g.* freshwater shrimp (*Gammarus pulex*)). Pathogens (*Myxoma* virus), invertebrates (*Psyllaephagus pilosus*) and vertebrates (*Ctenopharygdon idella*) have been successfully established as biocontrol agents in Europe.[6,7] Intentional introductions often result in widespread establishment of the aliens as a result of supplementary feeding, repeated introductions and the release of a high number of individuals. As might be expected, intentional introductions are responsible for the establishment in the wild of several alien species (Figure 1).

Although not intentional, the escape of alien plants and animals from managed environments into the wild is to be expected. Such "accidents" include the feralisation of crops (*e.g.* rape (*Brassica napus*)), livestock (*e.g.* goats (*Capra hircus*)) and farmed fish (*e.g.* Atlantic salmon (*Salmo salar*)); deliberate release from fur farms (*e.g.* American mink (*Mustela vison*)); escapes of ornamental animals (*e.g.* ruddy duck (*Oxyura jamaicensis*)) and plants (*e.g.* rhododendron (*Rhododendron ponticum*)); and the disposal of unwanted pets

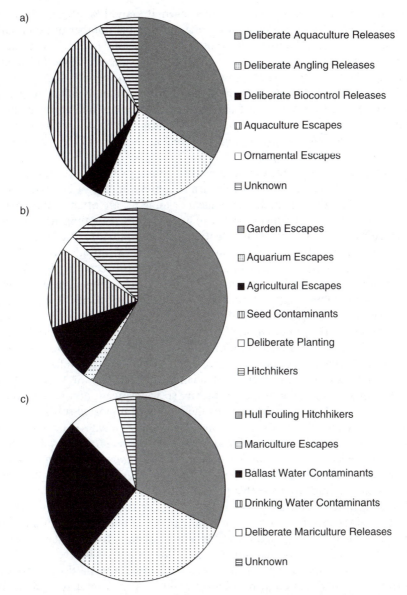

Figure 1   Major modes by which alien species have been introduced into European territory: a) freshwater species (data from the FAO Database on Introductions of Aquatic Species (DIAS)), b) terrestrial and freshwater plants in Scotland[31] and c) coastal marine fauna and flora in the British Isles.[55]

(*e.g.* red-eared slider (*Trachemys scripta elegans*)). Although relatively rare on an individual basis, the frequency of escapes is often high due to the large pool of potential escapees. Conservative estimates indicate that British gardens, plant centres and nurseries grow at least fifty times as many plant taxa as are found in the entire native flora.[8] Thus, even if only 1% of introductions establish, successful garden escapes represent a sizeable number of potentially problematic species. Some of the most pernicious and invasive non-native plants are the result of ornamental escapes (*e.g.* Japanese knotweed (*Fallopia japonica*), rhododendron and giant hogweed (*Heracleum mantegazzianum*)). Similarly, several of the most harmful non-native mammals introduced into Europe originated from escapes from fur farms: American mink, grey squirrel (*Sciurus carolinensis*), muskrat (*Ondatra zibethicus*), racoon (*Procyon lotor*), raccoon dog (*Nyctereutes procyonoides*), Canadian beaver (*Castor canadensis*) and coypu (*Myocastor coypu*).

Although less foreseeable than escapes of managed species, the introduction of aliens as contaminants of particular trade commodities is often predictable. Imported commodities (whether biological or mineral) are often contaminated with unwanted alien organisms. Alien pests, parasites or pathogens are often introduced with their hosts as unintentional "contaminants." In many cases, the impacts are restricted to either one alien host (*e.g.* Colorado beetle (*Leptinotarsa decemlineata*) on potato) or several hosts (*e.g.* tobacco whitefly (*Bemisia tabaci*) on ornamental plants). Of greater concern are pathogens that not only impact upon the imported host but also spread to native species. For example, spring viraemia (a serious viral disease) was introduced into Europe with ornamental carp (*Cyprinus carpio*) but, in addition to carp, wild species such as tench (*Tinca tinca*), roach (*Rutilus rutilus*) and pike (*Esox lucius*) are also vulnerable. Accidental contamination of grain supplies or feedstuffs presents a diverse route for the introduction of alien plant species into Europe. Seed contaminants are among some of the most widespread non-native weeds (Figure 1(b)). Although certification of the quality of imported seeds regulates this source of introduction into Europe, even today cereal seed samples are contaminated by alien crops (*e.g.* *Brassica* spp., *Daucus carota*) as well as non-native weed species (*e.g.* *Cerastium tomentosum*, *Lolium temulentum*).[8] While contamination is often under 1%, given the large numbers of seed sown each year this can amount to a sizeable pool of introductions.

The transport of soil and aggregates either for specific needs or as solid ballast provides a route of entry for a variety of plants and animals, often in the form of resilient resting stages such as seeds, spores or eggs. The range of potential species in imported soil is vast. For example, the New Zealand flatworm (*Arthurdendyus triangulatus*) was introduced into the British Isles in soil traded in pots, trays and root-balled plants.[9] Approximately 80% of the world's commodities are transported in ships, and at any given time approximately 35 000 ships are operating on the world's oceans.[10] These commodities are often transported on "one-way routes" with no suitable cargo available for the return trip. International cargo vessels are typically large (80–300 000 deadweight tonnage (DWT)), and may carry a third of their DWT as ballast when unladen. Ballast water (and the associated sediment) presents opportunities for the entry of microbes and the eggs, cysts and larvae of various species following discharge at recipient ports.[11] The life-histories of many marine species include planktonic stages that are small enough to pass through ballast water intake ports and pumps.[12] For example, almost 1000 species have been recorded in ballast water entering European ports and range from bacteria to 15-cm long fishes[13] (Figure 1(c)). The survival of the species in the ballast tanks usually shows a strong correlation with the length of the journey. This is especially pronounced for phytoplankton that require light for photosynthesis although species that produce resting spores are able to survive prolonged unfavourable conditions.[14]

The foregoing examples of alien species introductions are all in some extent directly related to the process of trade where the alien species is either the commodity, an alien contaminant of a commodity or deliberately transported material (*e.g.* ballast). However, alien species may be introduced independently of a traded commodity. Such species are described as hitchhikers and are closely linked to specific modes of travel and transport. Hitchhikers are possibly the least predictable of alien introductions and the only certainty is that increased trade and travel will facilitate further introductions. The most characterisable group of hitchhikers includes aquatic species that foul the hulls of boats and ships. A detailed survey of aliens introduced by shipping into the North Sea region revealed that although similar numbers of species are intro-duced through hull fouling (110), ballast water (110) and sediment (98), alien species were found in 96% of hull samples, but only 38% and 57% for ballast water and sediment, respectively.[15] The alien assemblage was primarily com-posed of crustaceans and bivalves. Thus the ballast tanks and associated structures may offer more opportunities for alien species than the ballast water or sediment, especially for surface-dwelling organisms. Nevertheless, the mech-anism by which many hitchhikers are introduced remains unknown. For example, the introduction of the South African narrow-leaved ragwort (*Senecio inaequidens*) is believed to have occurred via the soil carried on military equipment during the Second World War.[16] The plant is now spreading all over Europe along the road and the railroad systems. The horse-chestnut leafminer (*Cameraria ohridella*) was first observed in Macedonia in 1985 and has since spread throughout major parts of Central and Eastern Europe. While movement of ornamental planting material could exacerbate spread, most dispersal is undoubtedly through hitchhiking on cars, lorries and railway wagons. The almost ubiquitous occurrence of brown rats (*Rattus norvegicus*) as well as house mice (*Mus musculus*) on European islands is a testament to how these species have successfully hitchhiked along with humans.

Human activities have further facilitated long-distance dispersal of alien species through major infrastructural developments. The construction of canals has accelerated the transfer of species from one region to another. Canals are a feature of most European countries and they have their greatest impact when they connect two or more biogeographical areas that were previously isolated from each other. The Lessepsian invasion of Red Sea organisms through the Suez Canal has profoundly modified the ecosystem of the Eastern Mediterra-nean. The vast majority of migrational movement has been from the Red Sea to the Mediterranean and encompasses a wide range of taxa. The percentage of Lessepsian migrants in the Eastern Mediterranean is quite high: 7.1% of

Polychaeta, 22.9% of decapod crustaceans, 9.4% of Mollusca and 13.2% of fish.[17] The progressive interconnection of canals and rivers in central and eastern Europe has permitted the invasion of native species from the Caspian and Black Seas into the Baltic and North Seas.[18] The relatively low salinity in the Black and Caspian Seas enabled the survival of species within a freshwater canal network. The similarly low salinity in the Baltic Sea has led to this region supporting most of these Ponto-Caspian aliens.[19] Bridges and tunnels may play a similar role where they facilitate increased traffic between islands and mainland areas or connect neighbouring valleys in mountain areas.

3   Assessing the Pressure of Invasions on Ecosystems: Propagules, Pathways and People

On arrival following transport and subsequent release (whether intentional or unintentional) an alien species may or may not encounter a suitable environment. In Europe, the ecosystems most vulnerable to aliens are believed to be islands, lakes, rivers and in-shore marine areas.[20] The attributes that make particular ecosystems more susceptible to invasion have been scrutinised for several decades.[21,22] The only generalisations are that all European ecosystems are probably to some extent susceptible to invasion and that the greater the impact of human activity the more likely ecosystems are to be invaded.[23] Human activities have favoured alien taxa through the creation of new habitat (*e.g.* cities, harbours); annually ploughed fields, upland reservoirs or modification of existing habitats through changes in water availability; nutrient inputs, fire intensity and grazing pressure. High human population density and dense transport networks will increase the number and frequency with which individuals are introduced to an ecosystem especially as regards contaminants and hitchhikers.[24] However, alien species may be deliberately introduced even in relatively remote areas, such as the case of the European minnow (*Phoxinus phoxinus*) brought from Denmark and released as live bait into high elevation montane lakes of northern Norway.[25]

The likelihood of establishment in the wild will not only be a function of the susceptibility to invasion of the recipient ecosystem but also the degree to which it is exposed to different numbers of alien individuals. This is often described as the propagule pressure and refers to the number of potential viable reproductive units (*e.g.* seeds, spores, eggs, individual animals or plants) that are introduced to a specific ecosystem. Propagule pressure is often an important determinant of invasion but can be notoriously difficult to quantify. Data characterising introduction events are rare and almost restricted to information on deliberate releases of birds.[26] These data highlight that the volume and frequency of introduction events are the main pressures on ecosystems that affect alien establishment success. Propagule pressure will, in part, reflect the mechanism of entry into a new region. Examination of the alien flora of Scotland highlights significant differences in both the number and type of habitats colonised (Figure 2). Alien plant species deliberately released in large

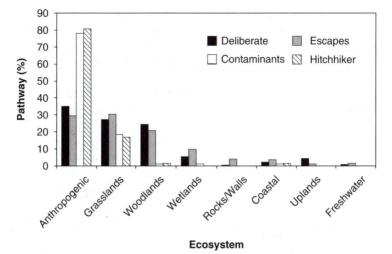

Figure 2   The occurrence of alien plant species in habitats of Scotland[31] in relation to mode of introduction (deliberate, escapes, contaminants and hitchhikers).

numbers for agriculture, silviculture or horticulture or that escaped from gardens are found in almost all habitats. In contrast, grain, ballast and soil contaminants or hitchhikers on other commodities such as timber are far less widespread. Indeed, a high proportion of contaminants and hitchhikers are found in anthropogenic rather than semi-natural ecosystems. Many of the seed contaminants are "convergent weeds", species that share many characteristics with the crop they contaminate. These taxa exist in highly managed and artificial environments and in the absence of soil disturbance tend to become locally extinct.[8]

The importance of propagule pressure is also seen in the marine environment where the proportion of alien taxa is higher in ports and estuaries than in other coastal areas or in the open seas.[27] Marine introductions are largely via contaminants or hitchhikers and this emphasises the orders of magnitude difference in the volume of international shipping entering European ports and estuaries compared to the scale of mariculture in the region. On a global scale some 3–5 billion tonnes of ballast water are transported annually.[10] Given that propagule pressure may play a key role in the susceptibility of European ecosystems to alien species, any attempt to relate invasions to key intrinsic ecosystem traits may prove difficult.

4   The State of the Union: Trends in the Distribution of Alien Species in Europe

As a first step in the assessment of biological invasions, several nations have begun to undertake audits of introduced species (*e.g.* Austria,[28] Ireland,[29]

Nordic countries,[30] Scotland,[31] England[32]). Unfortunately, comparison among different studies is hindered by variation in the degree to which marine, freshwater and terrestrial ecosystems were assessed and the range and quality of taxonomic detail. As a result, comparative examination of European trends is limited to relatively well-known taxa: higher plants, birds, fish and mammals. Plants represent an order of magnitude greater number of introduced species than any other taxonomic group but alien plant richness varies by an order of magnitude between France and Iceland (Figure 3). Alien freshwater fish similarly reveal a ten-fold variation between the least and most species rich countries (Iceland and France again) and have a higher average richness than

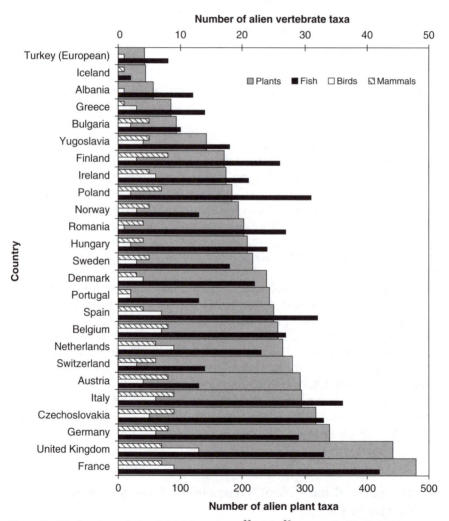

Figure 3   National trends in alien higher plants,[33] birds,[56] mammals (Societas Europaea Mammalogica, http://cgi.european-mammals.org) and fish[57] in 25 European states.

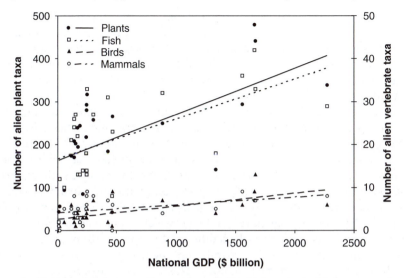

Figure 4   Relationship between the number of alien higher plants, birds, mammals and fish in 25 European states (same sources as Figure 3.3) and national GDP (from country reports in the CIA World Factbook, 2005: www.cia.gov/cia/publications/factbook/index.html). The correlations with GDP are significant for each taxon.

mammals. Alien mammal richness appears relatively similar across Europe. The number of alien taxa is correlated with national GDP for all four groups (Figure 4). This relationship probably reflects the important role of trade, consumerism and urbanisation in the introduction of alien species, their subsequent establishment and spread. Thus one legacy of the expansion of the European Union from 15 to 25 Member States in 2003 may be a future increase in the rate of alien invasions in Eastern Europe as their national GDP increases towards the European average. Not only do European nations differ in the total number of different alien taxa within their borders, but the species composition can also vary markedly. While several invasive taxa are widespread (*e.g.* the black rat (*Rattus rattus*), Japanese knotweed, Canada geese (*Branta canadensis*)), the distribution of species across countries is strongly skewed, with a few ubiquitous species and many species found in only one or two countries. For example, on average any one alien plant species occurs in fewer than four countries.[33] Thus the same species may be viewed very differently, even in neighbouring countries.

Although the majority of European nations have coastlines and thus jurisdiction over part of the marine environment, audits of marine alien species often take a regional perspective (*e.g.* Mediterranean, North-East Atlantic, Caspian, Baltic). A summary of much of this information for Europe has recently been published.[19] For the North Sea as a whole, about 80 alien taxa have been introduced.[27] The majority (almost 60%) are invertebrates (primarily crustaceans, molluscs, polychaetes and hydroids), introduced macroalgae comprise a further quarter (mainly red and brown algae), whilst protists account for 15%. Off the coast, approximately 6% of the macrobenthic species are alien while in estuaries it is almost 20%. Thus, national rather than regional marine inventories tend to reflect higher alien richness as a result of the greater influence of coastal and estuarine species leading to an underestimate of open water aliens, especially protists.

These brief examples of alien audits highlight several problems in adopting a pan-European approach to biological invasions. First, difficulties arise in the standardisation of the status of alien species. National studies often have access to far more detailed data, but classification of species may differ among countries. This is especially true in terms of the treatment of varieties, hybrids, reintroductions, translocations, feral species and naturally expanding populations. Guidelines for the classification of species status have only recently been suggested[34] (Table 1) and have yet to be widely implemented. Second, the heterogeneity in the degree to which different European nations are exposed to biological invasions may limit recognition of the risk that activities within their jurisdiction may pose to other nations. This is particularly important within Europe, with its shared coastline, trans-boundary mountain ranges and international watercourses, as species introduced into the territory of one nation can easily spread to its neighbours, sub-regions or the entire region. Third, species prioritised for management differ across Europe such that concerted actions should be planned at sub-regional scales. Finally, alien species in one European nation may be native in another. This poses considerable complexity on the development of regulations regarding trade within Europe.

5   Impacts on Biodiversity: Genes, Populations and Ecosystems

Alien species may impact on the populations of specific native species through hybridisation, by facilitating the spread of pathogens, via trophic impacts (grazing, predation, parasitism) and/or competition for resources. Reproductive isolation arising from geographic barriers such as mountain ranges, oceans and deserts is recognised as a major driving force in the evolution of species. When these geographic barriers are breached as a result of humans moving species across the globe, the genetic integrity of species may be threatened. Hybridisation between alien and native species is a potentially serious threat to biodiversity[35] (Table 2). Hybridisation may result in an infertile hybrid and this may lead to the decline of native species populations when hybrids represent the majority of offspring produced. Alternatively, the hybrids may be fertile and interbreed amongst themselves as well as the parental stock but generally perform less well than the native. Such "genetic pollution" threatens the integrity of native species and where this involves the spread of maladaptive genes, lower hybrid performance could lead to progressive native population declines. A further possibility is that the hybrid may exhibit new traits that enable it to occupy ecosystems from which either parent was previously absent or it may perform more vigorously in presently occupied ecosystems. While

Table 2   Known hybrids between alien and natives species in Europe and the consequence of hybrid offspring.

Taxon	Organism	Alien species	Native species	Consequence
Plants	Cordgrass	Spartina alterniflora	Spartina maritima	Allotetraploid hybrid is an aggressive invader of mudflats.
Amphibia	Frog	Rana ridibunda	Rana perezi	Hybridisation threatens genetic integrity of endemic native.
Fish	Whitefish	Coregonus peled	Coregonus lavaretus	Hybridisation threatens genetic integrity of endemic native.
Fish	Salmon/ Trout	Salmo salar	Salmo trutta	Infertile hybrid offspring reduce population growth rate of native.
Birds	Duck	Oxyura jamaicensis	Oxyura leucocephala	Hybridisation threatens genetic integrity of endemic native.
Mammals	Deer	Cervus nippon	Cervus elaphus	Hybridisation threatens genetic integrity of endemic native.
Mammals	Mink	Mustella vison	Mustella lutreola	Infertile hybrid offspring reduce population growth rate of native.

examples of each of these threats are known from Europe, a complete assessment is currently constrained by limited taxonomic knowledge, especially for invertebrates.

It may be expected that in the absence of physical, physiological or behavioural reproductive barriers, the likelihood of alien-native hybridisation will be a function of the genetic distance between two species. The greatest risks may therefore occur when subspecies are brought together. Some of the clearest cases relate to occasions where domesticated subspecies have established feral populations and interbreed with native subspecies. Examples include hybrids between domesticated and wild subspecies of goats (*Capra aegagrus hircus* and *C. aegagrus aegagrus*), cats (*Felis silvestris catus* and *F. silvestris lybica*), pigeons (*Columba livia domestica* and *C. livia livia*) and beet crops (*Beta vulgaris vulgaris* and *B. vulgaris maritima*). In certain cases, human use is sufficiently recent that no taxonomic distinction is made between domesticated and wild types (*e.g.* hybrids between farmed and wild salmon (*Salmo salar*)). Hybrids are also the result of humans moving different subspecies across Europe for commercial reasons. Mediterranean subspecies of the honeybee (*Apis mellifera*) and bumblebee (*Bombus terrestris*) have been introduced into north-western Europe where they hybridise freely with local subspecies.

While the expectation that subspecies might hybridise with one another is fairly predictable, hybrids at the species level are often harder to foresee. Why do hybrids occur in Europe among native and alien species of deer, ducks, frogs or oaks but not beavers, mink, rats or squirrels? Such taxonomic variation undeniably reflects the phylogenetic distances between species, chromosome number, degree of geographic separation, species mobility, whether the species are sympatric or parapatric and the character of reproductive isolation. For birds, although species occurring in sympatry were less likely to hybridise, hybrids were more likely when one of the species was reported as endangered and where hybrids were easier to detect.[36] For plants, the rather unsatisfying conclusion is simply the more alien species introduced within a particular genus the higher the number of hybrids produced.[37] Unfortunately for most taxa the number of known hybrids is insufficient to elucidate the relative importance of these factors and establish a predictive model of hybridisation risk.

The success of many alien species in new regions has been attributed to the escape from parasites and pathogens prevalent in their native ranges.[38,39] Yet, there are many cases where an alien species arrives with its parasites/pathogens and the latter have detrimental impacts on native species (Table 3). In some cases, the parasite or pathogen has a marked impact on native populations without unduly affecting the alien host. Dramatic examples in Europe include

Table 3   Examples of pathogens and parasites transmitted to native hosts following the introduction of specific alien species into Europe.

Taxon	Alien host	Native host	Alien parasite/pathogen
Plants	*Rhododendron ponticum*	*Quercus petraea*	Sudden oak death fungus
Crustacea	*Pacifastacus leniusculus*	*Austropotamobius pallipes*	Crayfish plague fungus
Insects	*Apis cerana*	*Apis mellifera*	Varroa mite
Fish	*Pseudorasbora parva*	*Leucaspius delineatus*	Rosette agent parasite
Fish	*Anguilla japonica*	*Anguilla anguilla*	Swim-bladder nematode
Mammals	*Cervus nippon*	*Cervus elaphus*	Asiatic blood nematode
Mammals	*Mustela vison*	*Mustela lutreola*	Aleutian disease virus
Mammals	*Sciurus carolinensis*	*Sciurus vulgaris*	Parapox virus

the transmission of parapox virus between alien grey and native red squirrels (*Sciurus vulgaris*) and plague fungus in North-American signal crayfish (*Pacifastacus leniusculus*) that has spread to native European crayfish (*Austropotamobius pallipes*). In these examples, the pathogen is believed to have facilitated the establishment and spread of the alien host. In other cases, the role of the parasite or pathogen in the spread of the alien host is less clear. For example, where the introduction of alien hosts has assisted the establishment of a parasite/pathogen but subsequently the latter has spread more widely via free-living stages (*e.g.* eel swim-bladder nematode) or several alternate native hosts (sudden oak death fungus). Often the impact of parasites and pathogens is most marked in commercial populations of hosts where densities are high. The wider impact on wild populations is more difficult to assess but can occur over a large spatial scale and long time period as illustrated by the decline of elms (*Ulmus procera*) in the UK following the introduction and spread of Dutch elm disease (*Ophiostoma ulmi*).

Where an alien predator has become successfully established it will more than likely subsist on a diet of native prey. Extrapolating from this observation to predictions regarding impacts on biodiversity is not straightforward in cases where predators have generalist feeding habits and impacts on any individual species may be limited. The racoon dog is an omnivorous predator but removal studies in Finland reveal only a limited impact on their avian prey.[40] However, impacts of generalist alien predators on specific native populations do occur. The American mink is held partially responsible for the decline in water vole populations (*Arvicola terrestris*) in the UK.[41] The muskrat preys, amongst other things, upon native freshwater mussels and can often lead to local population extinctions.[42] The introduction of an alien amphipod (*Gmelinoides fasciatus*) into eastern European lakes resulted in the extinction of native amphipods.[19] The predatory New Zealand flatworm is suspected of causing declines and local extinctions of earthworms in western Scotland.[9] The most marked predatory impacts are often found on islands where small populations of relatively naïve prey are exposed to food-limited alien predators (Table 4). In many cases the alien culprits are cats and rats and the victims are the flightless chicks of nesting seabirds.

Evidence of alien herbivores impacts on specific native plant species populations is largely drawn from the agriculture and forestry sector where introduced pests cause significant damage to crops and plantations. Outside of managed ecosystems, it is generalist vertebrate herbivores that have a reputation of negative impacts on biodiversity, especially on islands. Feral goats and to a lesser extent sheep and cattle have established populations on many islands as a result of deliberate introductions or escapes from domestic livestock. The most noticeable impacts have occurred where island plant communities evolved in the absence of ungulate (*i.e.* hoofed animal) grazing. Under such circumstances feral herbivores have modified entire island ecosystems.[43] For example, tussock grassland dominated by *Poa* spp. has been largely destroyed by feral ungulates on Tristan da Cunha, South Georgia, the Falklands, Kerguellen and New Amsterdam islands. This has resulted in a loss in plant and invertebrate

Table 4 The impacts of alien vertebrates on specific faunal elements of European island territories.[16,43,54]

Island	Territory	Region	Alien predator	Native prey
Ailsa Craig	UK	North Atlantic	Brown Rat	Manx Shearwater
Lundy	UK	North Atlantic	Brown Rat	Atlantic Puffin
South Uist	UK	North Atlantic	American Mink	Arctic Tern
Madeira	Portugal	North Atlantic	Brown Rat	Trocaz Pigeon
La Gomera	Spain	North Atlantic	Feral Cat	Giant Lizard
Tenerife	Spain	North Atlantic	Brown Rat	Laurel Pigeon
El Hierro	Spain	North Atlantic	Black Rat	Giant Lizard
Gran Canaria	Spain	North Atlantic	Feral Cat	Blue Chaffinch
Brittany	France	North Atlantic	Brown Rat	Rock Pipit
Swedish Isles	Sweden	Baltic	American Mink	Eider Duck
Bornholm	Denmark	Baltic	Brown Rat	Black-headed Gull
Baltic Islands	Finland	Baltic	American Mink	Black Guillemot
San Stephano	Italy	Mediterranean	Feral Cat	San Stephano Lizard
La Dragonera	Spain	Mediterranean	Black Rat	Balearic Shearwater
Corsica	France	Mediterranean	Black Rat	Cory's Shearwater
Capraia	Italy	Mediterranean	Feral Cat	Balearic Shearwater
Guadeloupe	France	Caribbean	Indian Mongoose	House Wren
Martinique	France	Caribbean	Indian Mongoose	Martinique Snake
Martinique	France	Caribbean	Indian Mongoose	Giant Lizard
St Kitts	UK	Caribbean	Green Monkey	St Kitts Bullfinch
Turks & Caicos	UK	Caribbean	Feral Cat	Ring-tailed Iguana
Falkland Islands	UK	South Atlantic	Brown Rat	Broadbilled Prion
Falkland Islands	UK	South Atlantic	Argentine Grey Fox	Upland Goose
South Georgia	UK	South Atlantic	Rat	Purple Martin
St Helena	UK	South Atlantic	Rat	Red-billed Tropicbird
Tristan da Cunha	UK	South Atlantic	Black Rat	Broadbilled Prion
Tristan da Cunha	UK	South Atlantic	Wekas	Diving Petrel
Tristan da Cunha	UK	South Atlantic	Feral Pig	Tristan Rail
Tristan da Cunha	UK	South Atlantic	Feral Cat	Tristan Thrush
La Reunion	France	Indian/Pacific	Feral Pig	Giant Tortoises
New Caledonia	France	Indian/Pacific	Feral Dog	Kagu
Pitcairn Island	UK	Indian/Pacific	Polynesian Rat	Murphy's Petrel
French Polynesia	France	Indian/Pacific	Black Rat	Tuamotu Sandpiper

species richness as well as increased soil erosion that has had a subsequent negative impact on ground nesting birds.[43] Pliny the Elder wrote in his *Natural History* that the invasion of rabbits (*Oryctolagus cuniculus*) on the Balearic Islands was such a severe problem that the help of Roman troops was sought to control them.[44] Rabbits continue to pose problems in the Canary Islands where they threaten unique plant communities[16] and the landscape of the British Isles has been much influenced by their grazing.[43] However, on islands off the coast of Britain, rabbit grazing has enriched the flora and rabbit burrows provide nest sites for seabirds.[43]

Anecdotal reports often suggest that alien taxa compete and displace native species. The larger, more aggressive Canadian beaver is believed to outcompete and replace the European beaver (*Castor fiber*) in northern Europe. Mandarin ducks (*Aix galericulata*) are assumed to compete with the native goldeneye

(*Bucephala clangul*) since both species nest in tree holes close to rivers and such sites are in limited supply. However, observational evidence for competition is usually insufficient and the most robust evidence stems from studies where the alien species has been removed experimentally. Such studies are few and far between. Removal of the invasive riparian weed Himalayan balsam (*Impatiens glandulifera*) highlighted the fact that the alien reduces plant diversity by about a third but that the plant species that respond most dramatically are other non-native plants.[45] In contrast, the experimental removal of the invasive seaweed (*Sargassum muticum*) from a semi-exposed rocky shore had a negligible effect on the low intertidal macroalgal assemblage.[46] Even with experimental studies, identifying the resources for which species might be competing can be challenging. Himalayan balsam may compete successfully for light with native species[45] but is also more attractive to insect pollinators and as a result reduces seed set in native plants where they co-occur.[47] But which form of competition is most important in determining impacts? There is a dearth of detailed experimental evidence for competition between alien and native species in Europe and solely anecdotal evidence should be treated with caution.

In many cases, the impact of alien species is to replace or reduce the abundance of ecologically equivalent native species and there are rarely wider ecological implications. However, in selected cases alien species may act as "ecosystem engineers" or "keystone species" leading to significant alterations in invaded ecosystems. Alien species that act as "ecosystem engineers" have the potential to transform ecosystems by altering underlying biogeochemical, hydrological and/or geomorphological processes. Wholesale ecosystem changes occur following colonisation of sand dunes by mimosas (*Acacia* spp.) that includes augmentation of soil nutrients, stabilisation of dunes and replacement of native plant species. Riparian habitats are prone to the impacts of alien burrowing animals such as the Chinese mitten crab (*Eriocheir sinensis*) and coypu (*Myocastor coypu*) that destabilise riverbanks and increase soil erosion as well as flood events. Dense populations of the freshwater Asiatic clam (*Corbicula fluminea*) may affect the structure of planktonic communities and thus shift primary production to benthic communities. Alien species may also have such a wide impact on the resident fauna and flora through competitive and trophic interactions that they are classed as "keystone species". One of the most pronounced shifts in ecosystems has been as a result of the recent invasion of the American comb jelly (*Mnemiopsis leidyi*) to the Black and Caspian Seas. This predatory ctenophore has led to significant declines in zooplankton abundance that subsequently reduced pelagic fish populations. In Spain, the Argentine ant (*Linepithema humile*) displaces not only native invertebrates but also vertebrates and even impacts on plants through disruption of myrmecochorous seed dispersal mutualisms.

In contrast to increasing information on the pathways of entry, distribution and abundance of alien species, current understanding of the impacts of alien species in Europe lags behind other areas of biological invasions. In the United States, invasive species cause major environmental damages and losses adding up to almost $120 billion per year and 42% of the species on the Threatened or

Endangered Species lists are at risk primarily because of alien invasive species.[48] These estimates of impact at a continental scale have undoubtedly helped raise the profile of alien species on the US political agenda. There is no figure for the environmental cost of alien species in Europe and the extent to which populations of Threatened or Endangered Species are impaired by alien species has not been quantified. Undoubtedly, part of the problem is the fragmentation of knowledge across different European nations and the absence of large-scale collaborative assessments of impacts across major biogeographic regions.[49] As a consequence, the number and impact of harmful invasive alien species in Europe is chronically underestimated, especially for species that do not damage agriculture or human health. Comparable estimates across Europe would play a pivotal role in informing policy and identifying resource priorities, yet to date these data are few and far between. In the absence of clear, quantitative messages regarding the current and future threats posed by alien species, it is difficult to see European policy evolving to meet the challenge of biological invasions.

6   Responding to the Threat of Biological Invasions: a European Policy Perspective

To combat the threat posed by biological invasions, several international policy instruments, guiding principles and procedures addressing alien species strategies are relevant to Europe (Table 5). Three broad areas of policy are germane to the prevention and management of biological invasions: a) organisms harmful to plants or plant products; b) animal and fish diseases; and c) species that may threaten wild fauna and flora.

For the first two categories, European states have a comprehensive framework of laws and procedures that are harmonised with international phytosanitary, zoosanitary and trade rules. Coverage is mainly focused on agricultural pests and diseases affecting crops, livestock and farmed fish. The framework provides for biosecurity controls in the form of certification, quarantine procedures and post-entry surveillance as necessary, as well as measures to control spread. The effectiveness of these instruments is facilitated through clearly identifiable host and pest targets as well as the direct economic benefits of regulation. Because these contaminant pests have economic impacts there exist strict regulations on imports. For example, the "Plant Pests" Directive of the European Union provides lists of pest species that must be banned from being introduced into particular Member States while the "Aquaculture" Directive legislates against the introduction of organisms pathogenic to aquaculture animals. Identification of target species facilitates the control and eradication of economic pests, pathogens and parasites. For deliberate releases, while European countries have different national regulations for the release of biocontrol agents, approval has often been based on European and Mediterranean Plant Protection Organisation (EPPO) regulations. In addition, EPPO has made a "positive list" of organisms employed in biological control,

Table 5   International conventions, agreements, directives and codes of conduct/guidelines concerning preventing the effects of introductions of alien species/organisms. The list is intended to be illustrative rather than definitive.[50]

Bern Convention on the Conservation of European Wildlife & Natural Habitat

Bonn Convention on Migratory Species of Wild Animals

European Community Directives:
Birds Directive
Habitats Directive

FAO Code of Conduct on Responsible Fisheries

International Conferences on the Protection of the North Sea:
Integration of Fisheries & Environmental Issues
Bergen Ministerial Declaration on the Protection of the North Sea

International Council for the Exploration of the Sea (ICES):
Code of Practice on the Introductions & Transfers of Marine Organisms

International Maritime Organisation (IMO):
Guidelines for the Control & Management of Ships' Ballast Water

International Plant Protection Convention

North Atlantic Salmon Conservation Organisation (NASCO):
Resolution to Minimise the Threats to Wild Salmon Stocks from Salmon Aquaculture
Resolution to Protect Wild Salmon Stocks from Introductions & Transfer
Guidelines for Action on Transgenic Salmon

OSPAR Commission for the Protection of the Marine Environment of the NE Atlantic:
Annex V of the OSPAR Convention

United Nations:
Convention on the Law of the Sea
Convention on Biological Diversity Convention on the Law of Non-navigational Uses of International Water Courses
Alien Species: Guiding Principles for the Prevention, Introduction & Mitigation of Impacts

The World Conservation Union (IUCN):
Guidelines for the Prevention of Biodiversity Loss Due to Biological Invasion

---

which include organisms known not to have negative side effects. These instruments highlight how policy established in the agricultural sector could be developed in relation to protection of the natural environment. However, the range of potential species targets is several orders of magnitude greater in semi-natural than managed ecosystems and the economic benefits are harder to quantify. Unfortunately, customs and quarantine practices developed to protect health and economic interests against diseases and pests have often been found to provide inadequate safeguards against species that threaten native biodiversity.[11]

All European states have ratified the Convention on Biological Diversity in which Article 8h recommends *"each Contracting Party shall, as far as possible and appropriate, prevent the introduction of, control or eradicate those alien species which threaten ecosystems, habitats or species"*. Most European states have a further commitment *"to strictly control the introduction of non-indigenous species"* (Bern Convention on the Conservation of European Wild-life and Natural Habitats) and both the "Habitats" and "Birds" Directives of the European Union also contain provisions to ensure alien introductions do not prejudice the local flora and fauna.[50] The Convention on International Trade in Endangered Species of Wild Fauna and Flora (CITES) enables nations to impose stricter controls on trade in certain species. It also establishes powers to restrict the introduction *"of live specimens of species for which it has been established that their introduction into the natural environment presents an ecological threat to wild species of fauna and flora"*. The legal, administrative and policy measures adopted by the European Union in sectors directly or indirectly concerned with invasive alien species are outlined in a specific thematic report to the Convention on Biological Diversity.[51]

Does this extensive policy background result in an effective strategy to combat invasive alien species? The answer depends on whether loopholes exist in legislation and if European states prioritise implementation on either a voluntary or a regulatory basis. Regrettably, one rather large loophole exists in that European legislation is constrained to: a) prevent deliberate rather than accidental introductions; b) exempt the major sources of accidental introductions from legislation (*e.g.* forestry and agriculture species, biocontrol agents, introductions into zoological and botanical gardens); and c) provide no commitment to eradicate or control established non-native species. Similarly, regulations under CITES may prevent import of invasive species (*e.g.* red-eared slider and American bullfrog (*Rana catesbeiana*)) in the European Community but has no jurisdictions on captive breeding and domestic sale by the pet industry. In theory, deliberate introductions and escapes should be the most straightforward to monitor and regulate but in practice developing legislation appropriate to the economic sectors responsible for such introductions has proved difficult. Although voluntary codes of practice have been promoted within the horticulture and pet trade, adoption of the polluter-pays principle, where the costs of recapture, eradication or control are allocated to the agent responsible for an unlawful introduction or escape, appears distant. For example, although fishing with live bait is illegal in Norway, it still occurs since anglers continue with traditional fishing techniques and tourists bring live bait from Sweden and Finland, where fishing with live bait remains legal.[25] Similarly, the International Council for the Exploration of the Sea (ICES) has developed a "Code of Practice" on the movement and translocations for fisheries and marine culture purposes to diminish the risks of detrimental effects from the intentional introduction and transfer of marine organisms, yet it has not always been closely followed.[27] Of greater concern is the fact that while deliberate introductions may in the future be regulated and controlled, at least to some degree, introductions of alien hitchhikers can be much harder to prevent, even with rigorous inspection and quarantine procedures. The International Maritime Organisation (IMO) has begun addressing this issue with regard to ballast water and has proposed ballast water exchange by ships in the open sea to flush out alien species, but this is not fully effective in removing organisms from ballast and may be subject to ship safety limits.[12]

Where preventive policies fail, strategic management must succeed in order to contain, eradicate or mitigate the threats posed by alien species.[52] However, such actions are frustrated by the degree to which European states prioritise invasive species management. An indication of the priority of tackling biological invasions can be gleaned from submissions under Article 8h of the Convention on Biological Diversity. As a group, the European states rate implementation of Article 8h as a significantly lower priority than do non-European nations. This difference between policy awareness and implementation results in insufficient resources being made available to target invasive species. Consequently, some impacts of invasions could have been reduced if European states had uniformly applied relevant codes of practice and taken rapid action to eradicate introduced species following their detection (*e.g.* grey squirrel invasion in Italy, Caulerpa (*Caulerpa taxifolia*) invasion in France, zebra mussel (*Dreissena polymorpha*) in the Baltic). Similarly, several biological invasions now threatening Europe might have been prevented by a higher level of awareness of invasive alien issues and a stronger commitment to address them (*e.g.* introduction of the comb jelly into the Aegean).

7   A Future Europe: Will Economic Integration Lead to Biotic Homogenisation?

Although there exists an increased awareness of the threat of invasive species to Europe, alien species continue to be introduced. Changes in both the motive and mode of introduction will shape future introductions into Europe and while regulation of deliberate introductions may reduce risk, the role played by hitchhikers should not be ignored. Nevertheless, alien species of European origin still constitute the major group and are a consequence of the importance of intra-European trade. More recently, between 1991 and 2001 the value of European trade doubled and with it so did opportunities for species introductions. Yet, in addition to this quantitative change in trade, a qualitative shift also occurred. Over this period, while the value of intra-European trade almost doubled, trade with China increased four-fold and trade with the ten European accession states increased over five-fold.[53] Thus not only is the frequency and volume of trade increasing but the sources of trade are becoming more diverse. Accordingly a greater diversity and frequency of contaminants and hitchhikers may be expected in the future. World Trade Organisation (WTO) proposals relating to the globalisation of trade often conflict with guidelines on invasive species management set up under the auspices of the Convention on Biological

Diversity.[25] Without closer attention from the WTO on international conventions and environmental agreements it will be difficult to manage invasion pathways. The drive to reduce the barriers to international trade is nowhere stronger than within the European Union. However, European economic sectors not only act as major sources of alien introductions but also suffer impacts from invasive species (Table 6). Thus, all sectors involved in activities

Table 6 Examples of situations where the major economic sectors act as sources of alien species in Europe and the problems that alien species (from all sources) cause within these particular sectors.

Sectors	Sector as source of alien taxa	Aliens as sector problems
Agriculture	Feral Crops: *Linum usitatissimum*	Agricultural Weeds: *Oxalis pes-caprae*
	Nectar/Pollen Sources: *Impatiens glandulifera*	Contaminated Seed: *Amaranthus retroflexus*
	Alien Pollinators: *Bombus* spp.	Hive Parasites: *Varroa destructor*
	Fur Farms: *Mustela vison*	Vertebrate Pests: *Nyctereutes procyonoides*
Aquaculture	Fish Stocking: *Salvelinus alpinus*	Alien Pathogens: Spring viraemia
Energy	Biomass Crops: *Miscanthus chinensis*	Cooling System Fouling: *Dreissena polymorpha*
Health	Medicinal Herbs: *Tanacetum parthenium*	Allergenic Pollen: *Ambrosia artemisifolia*
		Toxic Sap: *Heracleum mantegazzianum*
		Disease Vectors: *Rattus rattus*
Horticulture	Garden Plants: *Mimulus guttatus*	Garden Weeds: *Aegopodium podagraria*
	Landscaping: *Robinia pseudoacacia*	Urban Weeds: *Ailanthus altissima*
Industry	Imported Raw Materials: *Senecio squalidus*	Development Constraint: *Fallopia japonica*
	Pet Industry: *Trachemys scripta elegans*	
Mariculture	Mariculture escapes: *Crassostrea gigas*	Alien Parasites: *Mytilicola orientalis*
Silviculture	Plantation Exotics: *Pinus contorta*	Forestry Weeds: *Prunus serotina*
	Plantation Pests: *Anoplophora glabripennis*	Forestry Pests: *Sciurus carolinensis*
Tourism	Zoological Gardens: *Muntiacus reevesi*	Hybridisation with Natives: *Oxyura jamaicensis*
	Botanical Gardens: *Hedychium gardnerianum*	Biodiversity Loss: *Rhododendron ponticum*
	Sports Fishing: *Oncorhynchus mykiss*	Fishing Tackle Foulant: *Cercopagis pengoi*
	Game Introductions: *Sylvilagus floridensis*	
Transport	Ballast Water: *Eriocheir sinensis*	Hull fouling: *Caprella mutica*
		Air Strikes: *Branta canadensis*
Water	Freshwater transference: *Dreissena polymorpha*	Alien macrofoulants: *Corbicula fluminea*

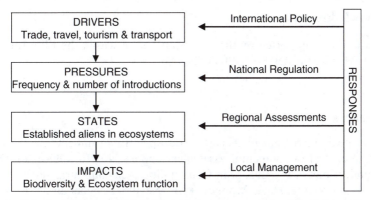

Figure 5   Schematic diagram highlighting the key components of the Drivers, Pressures, States, Impacts and Responses framework for biological invasions. Responses occur to each other component and reflect a hierarchy of scales from international policy to local management.

related to invasive species must have a role in implementing preventive and corrective action. (Figure 5)

The DPSIR framework can be adapted in a straightforward manner to characterise the threat to biodiversity from alien species. The framework thus presents a direct link between the economic drivers related to trade and the environmental consequences of invasion. The globalisation of trade is one of the primary drivers of biological invasions and establishes the regional pool of potential species introductions into any one region; internal trade, transport and urbanisation generate the pressure on natural ecosystems by increasing the propagule pressure. The interaction between the drivers and pressures is reflected in the state of the ecosystems, particularly the number of alien species established. Data on the impacts of the majority of alien species are limited but reflect a range of species and ecosystem effects. A hierarchical set of responses target each component. Driving forces are best addressed via international legislation that better reflects globalisation while national regulation and codes of practice can focus on specific pressures to selected ecosystems. Reporting on the states of ecosystems and species is probably best summarised at a regional scale while impacts are often managed for specific problems at a local level. The approach parallels the sequence of events associated with the invasion process where drivers determine the probability of species introduction into a region, pressures relate to patterns of establishments, states reflect population expansion and spread while impacts highlight invasive behaviour. The DPSIR framework highlights places where knowledge of different elements of biological invasions will be crucial to ensure productive dialogue between trade and conservation bodies and may help to target regulations and guidelines where the greatest risks exist. Only through such dialogue will Europe maximise economic wealth without sacrificing its environmental riches.

Acknowledgements

This study was supported by the European Union within the FP 6 Integrated Project ALARM (GOCE-CT-2003-506675).

References

1. European Environment Agency, "Europe's Environment: the Third Assessment", Environmental assessment report No. 10, Office for Official Publications of the European Communities, Luxembourg, 2003.
2. J. A. McNeeley, H. A. Mooney, L. E. Neville, P. Schei and J. K. Waage, "A Global Strategy on Invasive Alien Species", IUCN, Gland, Switzerland, 2001.
3. F. di Castri in "Biological Invasions: a Global Perspective", J. A. Drake, H. A. Mooney, F. diCastri, R. H. Groves, F. J. Kruger, M. Rejmánek and M. Williamson (eds), John Wiley & Sons, Chichester, 1989, 1.
4. A. W. Crosby, "Ecological Imperialism: The Ecological Expansion of Europe, 900–1900", Cambridge University Press, Cambridge, 1986.
5. P. Genovesi and C. Shine, "European Strategy on Invasive Alien Species", Council of Europe, Strasbourg, France, 2003.
6. A. J. Crivelli, *Biol. Cons.*, 1995, **72**, 311.
7. M. P. Chauzat, G. Purvis and R. Dunne, *Ann. Appl. Biol.*, 2002, **141**, 293.
8. P. E. Hulme in "Crop Science and Technology", British Crop Protection Council, 2005, 733.
9. B. Boag and G. W. Yeates, *Ecol. Appl.*, 2001, **11**, 1276.
10. J. T. Carlton, in "Invasive Species and Biodiversity Management", O. T. Sandlund, P. J. Schei and Å. Viken (eds), Kluwer Academic Publishers, Dordrecht, 1999, 195.
11. IUCN, "Guidelines for the Prevention of Biodiversity Loss Caused by Alien Invasive Species", IUCN, Gland, Switzerland, 2000.
12. IMO, Resolution A. 868, 29, 1997.
13. S. Gollasch, H. Rosenthal, H. Botnen, M. Crncevic, M. Gilbert, J. Hamer, N. Hülsmann, C. Mauro, L. McCann, D. Minchin, B. Öztürk, M. Robertson, C. Sutton and M. C. Villac, *Biol. Inv.*, 2003, **5**, 365.
14. G. M. Hallegraeff and C. X. J. Bolch, *J. Planct. Res.*, 1992, **14**, 1067.
15. S. Gollasch, *Biofouling*, 2002, **18**, 105.
16. European Commission, "Alien Species and Nature Conservation in the EU. The Role of the LIFE Program", Office for Official Publications of the European Communities, Luxembourg, 2004.
17. B. Galil, *Biol. Inv.*, 2000, **2**, 177.
18. A. Bij de Vaate, K. Jazdzewski, H. A. M. Ketelaars, S. Gollasch and G. Van der Velde, *Can. J. Fish. Aqu. Sci.*, 2002, **59**, 1159.
19. E. Leppakoski, G. Gollasch and S. Olenin, "Invasive Aquatic Species of Europe. Distribution, Impacts and Management", Kluwer Academic, Dordrecht, 2002.

20. V. Heywood, "Global Biodiversity Assessment", Cambridge University Press, Cambridge, 1995.
21. M. J. Crawley, in "Colonisation, Succession and Stability", M. J. Crawley, P. J. Edwards and A. J. Gray (eds), Blackwell Scientific Publications, Oxford, 1987, 429.
22. M. Rejmánek, in "Invasive Species and Biodiversity Management", O. T. Sandlund, P. J. Schei and Å. Viken (eds), Kluwer Academic Publishers, Dordrecht, 1999, 79.
23. P. E. Hulme, *Oryx*, 2003, **37**, 178.
24. P. E. Hulme, in "Island Ecology", J. M. Fernandez Palacios (ed.), Asociación Española de Ecología Terrestre, La Laguna, Spain, 2004, 337.
25. B. -Å. Tømmerås, A. Jelmert, T. Rafoss, L. Sundheim, F. Ødegaard and B. Økland, "Globalisation and Invasive Alien Species", Norwegian Institute for Nature Research, 2001.
26. T. M. Blackburn and R. P. Duncan, *Nature*, 2001, **414**, 195.
27. K. Reise, S. Gollasch and W. J. Wolff, *Helgoländer Meeresunters*, 1999, **52**, 219.
28. F. Essl and W. Rabitsch, "Neobiota in Österreich", Umweltbundesamt, Vienna, Austria, 2002.
29. K. Stokes, K. O'Neill and R. A. McDonald, "Invasive Species in Ireland", Unpublished report to Environment & Heritage Service and National Parks & Wildlife Service, Quercus, Queens University Belfast, Belfast. 2004.
30. I. Weidema, *Nord*, 2000, **13**.
31. D. Welch, D. N. Carss, J. Gornall, S. J. Manchester, M. Marquiss, C. D. Preston, M. G. Telfer, H. Arnold and J. Holbrook, "An Audit of Alien Species in Scotland", Scottish Natural Heritage Review, Edinburgh, 2001, 139.
32. M. Hill, R. Baker, G. Broad, P. J. Chandler, G. H. Copp, J. Ellis, D. Jones, C. Hoyland, I. Laing, M. Longshaw, N. Moore, D. Parrott, D. Pearman, C. Preston, R. M. Smith and R. Waters, *Engl. Nat. Res. Rep.*, 2005, **662**, 1.
33. E. F. Weber, *J. Veg. Sci.*, 1997, **8**, 565.
34. D. M. Richardson, P. Pysek, M. Rejmanek, M. G. Barbour, F. D. Panetta and C. J. West, *Divers. Distrib.*, 2000, **6**, 93.
35. N. C. Ellstrand and K. A. Schierenbeck, *Proc. Natl. Acad. Sci. USA*, 2000, **97**, 7043.
36. C. Randler, *Ibis*, 2006, **148**, 459.
37. C. C. Daehler and D. A. Carino, *Biol. Inv.*, 2000, **2**, 93.
38. C. E. Mitchell and A. G. Power, *Nature*, 2003, **421**, 625.
39. M. E. Torchin, K. D. Lafferty, A. P. Dobson, V. J. McKenzie and A. M. Kuris, *Nature*, 2003, **412**, 628.
40. K. Kauhala, *Folia Zool.*, 2004, **53**, 367.
41. S. P. Rushton, G. W. Barreto, R. M. Cormack, D. W. Macdonald and R. Fuller, *J. Appl. Ecol.*, 2000, **37**, 475.
42. J. Jokela and P. Mutikainen, *Can. J. Zool.*, 1995, **73**, 1085.
43. C. Lever, "Naturalized Animals", Poyser Natural History, London, 1994.

44. J. Clutton-Brock, "A Natural History of Domesticated Mammals", Cambridge University Press, Cambridge, 1999.
45. P. E. Hulme and E. T. Bremner, *J. Appl. Ecol.*, 2006, **43**, 43.
46. I. Sanchez and C. Fernandez, *J. Phycol.*, 2005, **41**, 923.
47. L. Chittka and S. Schürkens, *Nature*, 2001, **411**, 653.
48. D. Pimentel, R. Zuniga and D. Morrison, *Ecol. Econ.*, 2005, **52**, 273.
49. M. Vilà, M. Tessier, C. M. Suehs, G. Brundu, L. Carta, A. Galinidis, P. Lambdon, M. Manca, F. Medail, E. Moragues, A. Traveset, A. Y. Troumbis and P. E. Hulme, *J. Biogeog.*, 2006, **33**, 853.
50. C. Shine, N. Williams and L. Gundling, "A Guide to Designing Legal and Institutional Frameworks on Alien Invasive Species", IUCN, Gland, Switzerland, 2000.
51. European Commission, "Thematic Report on Alien Invasive Species", Second Report of the European Community to the Conference of the Parties of the Convention on Biological Diversity, 2003.
52. P. E. Hulme, *J. Appl. Ecol.*, 2006, **43**, 835.
53. European Commission "Panorama du Commerce de l'Union Européenne", Office for Official Publications of the European Communities, Luxembourg, 2004.
54. F. Courchamp, J. -L. Chapuis and M. Pascal, *Biol. Rev.*, 2003, **78**, 347.
55. N. C. Eno, R. A. Clark and W. G. Sanderson, "Non-native Marine Species in British Waters: a Review and Directory", Joint Nature Conservation Committee, Peterborough, 1997.
56. W. J. M. Hagemeijer and M. J. Blair, "The EBCC Atlas of European Breeding Birds: Their Distribution and Abundance", T. & A.D. Poyser, London, 2002.
57. B. Elvira, Council of Europe, T-PVS, 2001, 6.

The Deep Sea: If We Do Not Understand the Biodiversity, Can We Assess the Threat?

PAUL TYLER

1 Introduction

The first indication, other than mythological, that animals occurred in the deep sea was the serendipitous recovery of the euryalid brittlestar *Astrophyton linki* on a sounding line cast to 1800 m in the Baffin Sea in 1818.[1] The early 19th century saw an increasing interest in marine organisms in shallow water and fisheries were well established over the continental shelves. Towards the middle of the century, two groups started showing interest in animals collected from deeper depths. Edward Forbes[2] was sampling in the Aegean Sea and recorded that below 600 m there was little or no life, a concept that became known as the "azoic" theory. At the same time G. O. and M. Sars, father and son, were sampling in deep Norwegian fjords, finding rich communities of what we now call megabenthos, demonstrating that animals had the ability to live and thrive in waters of considerable depth.[3] Not long afterwards, individuals of the Royal Society (UK) began to collaborate with the Admiralty and the cruise of HMS *Lightning* could be considered the first oceanographic cruise to deep water. Although not primarily biological, the cruise demonstrated the presence of a ridge, between Scotland and the Faroes, separating the Norwegian Sea from the main North Atlantic. North of what became known as the Wyville-Thomson Ridge (now the Scotland-Faroes-Iceland-Greenland Ridge) was known as the "cold" area (temperatures < 5°C), whilst to the south was the "warm" area (temperatures 4.5 to 8.5°C). In the subsequent two years, HMS *Porcupine* was to sample this area, as well as to the west and south-west of Ireland and the western Mediterranean[4] (see Figure 1). Even at the deepest stations sampled, the nets brought up large marine organisms. The "azoic" theory was dead.

Issues in Environmental Science and Technology, No. 25
Biodiversity Under Threat
Edited by RE Hester and RM Harrison

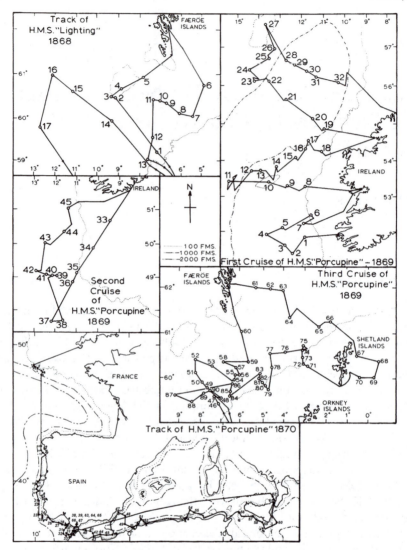

Figure 1 Tracks of HMS *Lightning* and *Porcupine* in the NE Atlantic and Mediterranean (1868–1870) establishing the presence of fauna in deep water.[4]

The driving force behind these expeditions was Charles Wyville-Thomson and it was his energy that led to the formulation of what became known as the *Challenger* Expedition. HMS *Challenger* was a 60 m, 2300 ton displacement ship fitted with steam winches. Between 1872 and 1876 she circumnavigated the world sampling some 362 observing stations and collecting a wide variety of organisms from the greatest depths (see Figure 2). This cruise, *inter alia*, demonstrated that animals could live at the greatest depths and the deep sea appeared to be a refuge for taxa such as the stalked crinoids thought to be long

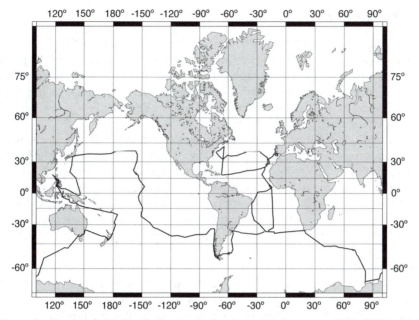

Figure 2 Track of HMS *Challenger* during her global voyage 1872–1876, which established the existence of deep-sea fauna.

extinct. The publication of this material fell to the organisation (and financial cost) of John Murray who published the *Challenger Reports* over the next 30 years covering every taxon that had been collected. The *Challenger Reports* can be described as the first chronicle of deep-sea biodiversity.

Interest in the organisms found in the deep sea exploded with all the developing nations mounting deep-sea expeditions. This has become known as the 'heroic age' of deep-sea exploration with *inter alia* the *Albatross* and the *Blake* (US) in the western North Atlantic, the Caribbean and Gulf of Mexico, the various ships of the Prince of Monaco in the Mediterranean and eastern Atlantic, the *Ingolf* (Denmark) and *Michael Sars* cruise in the North Atlantic, the *Mabahiss* (Anglo-Egyptian) in the Indian Ocean and the *Siboga* (Sweden), worldwide.[3] The culmination of this heroic age was the Danish *Galathea* expedition of 1950–1952 that sampled specifically the deepest trenches of the world ocean and showed that animals were capable of living in the very deepest part of the world ocean. Although the many expeditions over this 75-year period collected numerous new species, the equipment used to collect consisted mainly of coarse mesh trawls, tangles and fish traps, and virtually none of it retrieved quantitative samples. As a result the megafauna became well known and it was assumed that biodiversity was generally low in the deep sea. It was to be with new equipment and new vigour that the true contributors to biodiversity in the deep sea were to be recognised in the 1960s.

2   The Deep Sea

The ocean covers two-thirds of the surface of earth. However, if we examine a hypsographic curve (Figure 3) we will see that 50% of the surface of the Earth lies between 3000 and 6000 m depth. Thus the deep-sea environment, in terms of area, is the largest on Earth. The area of the deep-sea floor has been estimated at 3×10^8 km^2, whilst the volume of the deep ocean, *i.e.* the deep-sea pelagic environment, is 1.4×10^9 km^3.

The traditional concept of the deep sea is that it is a tranquil, unchanging environment of slow physical and biological processes. This was certainly the perception up to the 1970s, as deep-water temperatures are low (with notable exceptions) and salinity remarkably constant[5] but we now know that the deep-sea floor, in particular, is a very heterogeneous environment. If we could walk along the seabed from the continental shelf down into the abyssal plains we would pass over the continental slope cut in many places by canyons and, in others, covered by massive deep-water coral reefs (Figure 4A). In other areas the sediment cover appears to provide a barren lifeless environment (Figure 4B). On the steep parts bedrock may be exposed with no more than a dusting of sediment (Figure 4C). Nearing the base of the slope it is possible to find the remnants of sediment slumps and slides and the results of turbidity currents. Turbidity currents are not the only high-energy events occurring in the deep sea. In areas of strong surface vorticity, energy can be imparted downwards from the surface to the deep-sea bed and under certain conditions will cause "benthic storms" where the normal slow currents are reversed and accelerated and erode massive amounts of sediment.[6] As the energy input declines sediment is deposited in thick layers on the seabed. The fauna in such regions appears to be adapted to such variation in energy input.[5,7]

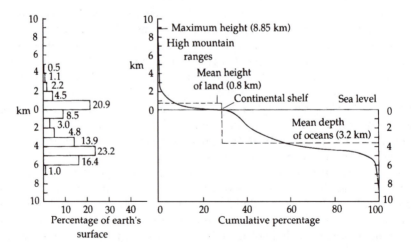

Figure 3   Hypsographic curve showing the relative proportions of the seabed at different depths. 50% of the Earth's surface lies between 3000 and 6000 m depth.[10]

Figure 4   Examples of different environments in the deep sea. **A**. Highly biodiverse
deep-water corals; **B**. Typical bathyal/abyssal sedimentary environment with
high biodiversity of small macrofauna; **C**. Fauna on steep rocky surface also
can show high diversity; **D**. Hydrothermal vent environment with high
biomass and relatively low diversity.

Stretching out from the base of the continental slope are the extensive abyssal
plains that are generally sediment covered and show very little relief. However,
arising from many abyssal plains, especially in the Pacific, are seamounts that
are often steep sided and extend for up to thousands of metres above the
surrounding plain.[8] As with all objects placed in a flow, they modify the flow
and in turn different fauna inhabit different parts of the seamount. Flow varies
around the seamount as a result of Coriolis force and thus organisms at the
same level but on different sides of the seamount may be subject to different
flow regimes[9].

As we approach the centre of the main oceans the sediment drape gets
thinner and the bare rocks of the mid-ocean ridge start to show through.[10] The
mid-ocean ridge (MOR) is where the new oceanic plate is formed that causes

the seafloor to spread. It was in 1977 that hydrothermal vents were discovered along the Galapagos Spreading Ridge, which were ultimately shown to have their own unique faunas (Figure 4D), quite different from those of the surrounding rock and sedimentary areas.[11]

At the far end of the oceanic spreading plates is the subduction zone, best exemplified by the so-called "ring of fire" in the Pacific. In subduction zones the oceanic plate is subducted beneath the lighter continental plate and at this point there is the formation of the deep-ocean trenches, the deepest being the "Challenger" Deep in the Marianas Trench at a little over 11 000 m depth. Subduction zones are generally areas of geological instability and the fauna is adapted to this instability.

Thus we see that the deep-sea bed throughout the world ocean is highly heterogeneous, although the mosaic of environments that make up this hetero-geneity vary in scale from patches of deep-water reefs covering a few square metres to the extensive abyssal plains that cover thousands of square kilome-tres. Although this heterogeneity is important in structuring biodiversity in the deep sea, superimposed on this physical environment is energy input. The deep sea (with some notable exceptions) is a heterotrophic environment requiring an input of organic matter from surface production to maintain it. Conceptualised models of this input vary from large food falls such as whales, wood and large macrophytes to the material we now call "phytodetritus", the product of the breakdown and sinking of phytoplankton production in surface waters.[12] This surface production varies considerably, being low in the centre of the tropical oceanic gyres, highly seasonal at temperate latitudes and higher still in terms of total production in regions of upwelling. In the case of regions of upwelling production can be so high that detritus sinks to bathyal depths where bacterial heterotrophic processes are so active they reduce the oxygen content of the water column to such an extent that the water column over the seabed is hypoxic ($< 0.2 \, \mathrm{ml} \, \mathrm{l}^{-1}$) or even anoxic as below 200 m in the Black Sea (see ref. 13 for review). This hypoxia has noticeable consequences for biodiversity in these regions.

It is within this context of the physico-chemical background of the vast expanses of the deep sea that we have to examine biodiversity. Because of the areal extent of the deep sea only small areas have been considered in detail and then these data extrapolated to the total ocean – with highly debateable results.

3   Understanding Modern Deep-sea Biodiversity

The quantitative approach to understanding deep-sea biodiversity has its origins in the 1960s with Howard Sanders and Bob Hessler, both then at the Woods Hole Oceanographic Institution outside Boston. Together with George Hampson, these biologists developed the anchor dredge, subsequently the anchor box dredge, and the epibenthic sled and deployed them along a series of stations that became known as the "Gay Head-Bermuda Transect" from the continental shelf down to 5000 m.[14] The significance of these samplers is that they, for the first time, retained the smallest individuals of the macrobenthic community. Careful sieving of the samples on deck and subsequent sorting of the residue in the lab revealed an astonishingly high biodiversity, composed not of megabenthos but of those organisms retained on a 0.42-mm sieve.[15] Dominant amongst this smaller fauna were polychaetes, peracarids and small molluscs.[14] Such a discovery was the first recognition that the deep sea may be one of the main repositories of biodiversity on the planet.

This practical demonstration of high biodiversity in the deep sea resulted in a series of theories about why the deep sea was so diverse. These included equilibrium theories such as the stability-time hypothesis,[16] which predicted high biodiversity when, under stable conditions, biological interaction resulted in a wide range of adaptations, and non-equilibrium theories such as the "intermediate disturbance hypothesis",[17] which predicted that peak diversity would be found at intermediate levels of disturbance.[10]

In the following years there were several programmes that examined the biodiversity of the deep sea with varying conclusions as to exactly how diverse the deep sea was. The main consensus was that there was considerable variability.

The next major impact was a seminal paper by Grassle and Maciolek.[18] The introduction of the USNEL box core[19] had provided deep-sea biologists with a truly quantitative apparatus for deep-sea sediment sampling. Grassle and Maciolek took 233 "vegematic" box cores (see ref. 10) along a 176-km transect between 1500 and 2500 m depth on the continental slope off New Jersey and Delaware. Samples were taken in different seasons. The analysis revealed 798 species (from a total sampled area of 21 m^2 by using only the inner 9 subcores (of 25 in total)) from these samples, with single individuals of species being highly represented. To this point there was no controversy. Grassle and Maciolek then extrapolated their data up to the deep-sea global ocean. Assuming depths below 1000 m account for 3×10^8 km^2 of seabed and that one species could be added for each 1 km^2, a total of 10^8 species would be found in the deep sea. This figure was refined to 10^7 as the authors recognised that oligotrophic and very deep water would contain fewer species. Grassle and Maciolek suggested that this estimate may even be conservative as species accumulation across contours is greater than along contours (see ref. 20). Such diversity was a result of microhabitat heterogeneity, with few barriers to dispersal, disturbance created by feeding of larger animals and patchy food resources.

The response was swift and highly critical. May,[21] whilst accepting Grassle and Maciolek's data, questioned their extrapolated estimate of 10^8 species in the deep sea. If ease of immigration and long-distance dispersal are important in an environment without apparent barriers it is not justified to extrapolate from a local rate of species addition. May also questioned the proportion of new species identified by Grassle and Maciolek and proposed that the total number of benthic species is about twice that already known. May thus concluded that the total number of species in the deep sea is "unlikely to exceed 500 000".[21]

Conversely, John Lambshead of the Natural History Museum, London, suggested there may be as many as 10^8 nematode species alone in the deep sea. The term "hyperdiversity" entered the deep-sea terminology.[22] To understand some of the reasoning in Grassle and Maciolek's analysis we have to examine variations in biodiversity in the deep sea.

4 Patterns of Biodiversity in the Deep Sea: Benthos

4.1 *With Depth*

Zonation has long been recognised as a pattern of changing species composition with depth.[10] Mike Rex's group in Boston have been foremost in determining the overall variation of biodiversity with depth (reviewed by ref. 23). Rex[24,25] was the first to plot total diversity within individual taxa with depth (Figure 5). Rex used the Gay Head-Bermuda Transect data for gastropods, protobranch bivalves, polychaetes and cumaceans. In addition, he plotted total megabenthos and total fish. The data suggested there was an increase in biodiversity down the slope to depths of between 1200 m and 3000 m below

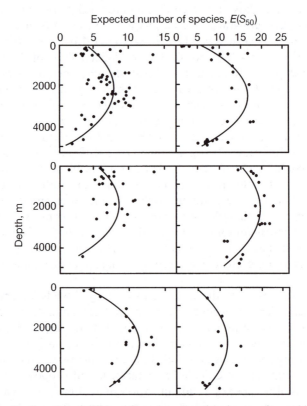

Figure 5 Species diversity of different faunal groups showing maximum biodiversity at bathyal depths.[24]

which diversity decreased with depth to 5000 m. However, as Rex and his co-workers recognised, such a pattern may be a result of scale and the density of organisms decreasing with depth.

The use of modern molecular methods has started to assist in clarifying diversity with depth.[26,27] Depth is more important than horizontal distance in structuring populations of the deep-sea protobranch *Deminucula atacellana*.[27] Populations from the North American, West European and Argentine Basins were well differentiated, but within individual basins populations at the same depths but over 1000 km apart were more similar than populations separated by 100 m depth. Such structuring with depth is seen in deep-sea asteroids,[28] with depth separation resulting in the recognition of separate species.

The causes of such a pattern are not easy to resolve. Stuart *et al.*[23] proposed the dynamic equilibrium model[29] as the best explanation. This proposes that superior competitors will out-compete inferior species and thus decrease diversity, whereas disturbance inhibits a community approaching equilibrium, reduces exclusion, promotes coexistence and thus increases diversity. In addition, one of the main structuring elements of deep-sea biodiversity is sediment variation with depth.[23] As many (if not most) deep-sea macrofaunal species are deposit feeders, macrofaunal diversity increases with a decrease in sediment grain size.[30] A "source-sink hypothesis" was proposed for biodiversity at abyssal depths in relation to bathyal depths.[31] At abyssal depths individual density is low and there is an "Allee effect", where sperm from a male is too dilute if transported even over a few metres and will not fertilise eggs, therefore making the populations unsustainable. Thus the populations of individual species are not viable over the long term and have to be replenished from populations reproducing at bathyal depths. Accordingly there is a selection against certain species in the abyss and biodiversity is reduced. Abyssal populations are "regulated by a balance between chronic extinction arising from vulnerability to Allee effects and immigration from bathyal sources".[31]

The pattern of mid-bathyal maximum in diversity is not universal in the deep sea. In areas where water column oxygen deficiency impinges on the seabed, such as under regions of upwelling, diversity decreases (although biomass often increases) at mid-depths (see ref. 32). In addition, sinking organic matter passes through the oxygen minimum zone with little or no heterotrophic recycling and as a result there is often an injection of organic matter into deeper layers of the ocean under oxygen minimum zones than under "normal" oxic conditions. As a result the depth of maximum diversity is deepened.

4.2 With Latitude

Latitudinal decline in biodiversity is well established in terrestrial populations, shallow marine and pelagic populations.[33] However, it was not until 1993 that such patterns were recognised in deep-sea populations. In a detailed analysis by Rex *et al.*[34] (Figure 6) a decline in biodiversity (isopods, gastropods, bivalves combined) was seen in deep-sea communities with increase in latitude. This was

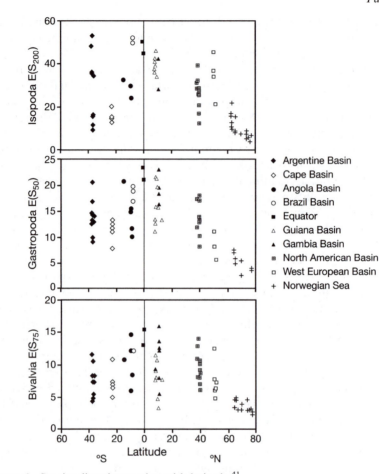

Figure 6 Species diversity varying with latitude.[41]

especially obvious in the northern Atlantic where biodiversity reached a minimum in the Norwegian Sea. Although criticised because the low values of the Norwegian Sea represented the effects of the last ice age, there is still a significant decrease in biodiversity with latitude if the Norwegian Sea is excluded.[23] A similar pattern was recognised in the South Atlantic although it was not as obvious because of the high species diversity observed in the Argentine Basin.[23,34] However, a more pronounced decline with latitude was observed when considering the foraminifera.[35]

Lambshead *et al.*[36] reported a latitudinal decline in nematode diversity with increasing latitude from 13 to 56°N in the North Atlantic, which they related to decrease in surface productivity. However, Rex *et al.*[37] questioned Lambshead *et al.*'s analysis as they had not corrected for depth. When corrected for depth, latitude only accounted for 8% of the biodiversity. Such are the difficulties of interpreting biodiversity in such a large 3-dimensional environment!

One of the biggest problems is comparability between sampling programmes. Biodiversity of isopods, gastropods and bivalves in the deep Weddell Sea (Antarctica) were similar to the deep tropical Atlantic and thus there was no latitudinal gradient.[38] However, deep-sea biodiversity in Antarctic waters is still imperfectly understood. Species diversity in the deep Weddell Sea and South Sandwich forearc basin is nowhere near asymptotic on rarefaction curves and thus is severely undersampled.[39] In a review of Antarctic deep-sea biodiversity in relation to the rest of the global ocean, Gage[40] suggested that basin confluence between the Antarctic and Atlantic, Indian and Pacific deep basins may have encouraged dispersion of species both into and out of the Antarctic basins, resulting in only superficial regional identity. Caution is urged because of the variability in sampling and analytical methodologies when examining large-scale geographic variation.[41]

4.3 With Productivity

Gradients in productivity were predicted to influence species diversity.[42] As a broad rule, biodiversity within a single taxon increases from low to intermediate productivity and declines with higher productivity.[43] As the deep sea relies almost exclusively on the rain of organic matter from surface production there is generally a decrease in productivity and flux as one moves away from the coast to the waters of the oligotrophic gyres, especially along well-oxygenated margins. The gradient is complicated by the increase in water depth over the same scale. Other complications are the increase in surface productivity along the equatorial belt in the central Pacific where higher surface production occurs because of waters advected from the Californian and Peru upwellings.

An opportunity to study latitudinal- and productivity-based variation in species diversity arose with the establishment of a north–south transect just to the north of the equator in the central Pacific at depths between 4300 and 5100 m.[44] This transect sampled stations at the equator, 2°N, 5°N, 9°N and the Hawaiian Ocean Time Series (HOTS) station at 23°N. Along this latitudinal gradient surface productivity decreases northward, towards the centre of the North Pacific Gyre. At the equator there is equatorial upwelling with an annual surface productivity of $230\,g\,C\,m^{-2}\,y^{-1}$ and a flux to 2000 m of $1.6\,g\,C\,m^{-2}\,y^{-1}$. This decreases to $105\,g\,C\,m^{-2}\,y^{-1}$ and $0.4\,g\,C\,m^{-2}\,y^{-1}$, respectively, at the HOTS station.[45] Quantitative sampling using box cores was undertaken using an USNEL Box corer or a Barnett Multiple corer.

The data for species diversity were equivocal. A clear latitudinal gradient in species richness of the nematode fauna was demonstrated, decreasing as one moves north from the equator[46] (Figure 7). It was concluded that there was a positive correlation between species richness and the organic flux to the seabed. By contrast, only weak evidence of a monotonic increase in polychaete diversity with surface production over the same gradient and rates of species turnover were slow.[47] This dichotomy was noted[46] and explained by the relatively low ability of nematodes to disperse, compared to other deep-sea benthic taxa, and

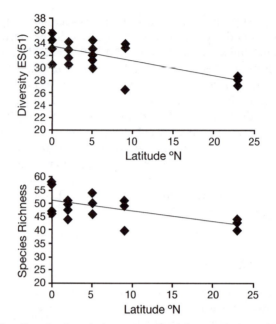

Figure 7 Species diversity in relation to productivity moving northwards from the equator in the central Pacific.[46]

possibly their resistance to low oxygen conditions, similar to those that occurred in the Palaeocene and Eocene. As this time, the hypothesis of a latitudinal gradient in diversity associated with a latitudinal variation in flux from surface primary production remains in need of further testing.

The difficulty of relating species diversity to productivity is thoughtfully addressed in a review of regional diversity in the deep sea.[43] The data gathered showed that diversity could vary positively, negatively or unimodally with productivity, depending on where the sampling site fell on the diversity/ productivity curve (Figure 8). In the central Pacific and in the NE Atlantic the sites may fall on the ascending part of the curve. Regions experiencing high organic carbon input, *e.g.* the North Carolina slope, may be found on the descending part of the diversity/productivity curve. The main problem is that there are many compounding factors and the relationship between diversity and productivity needs to be better substantiated.[43]

4.4 Hydrothermal Vents, Cold Seeps and Whale Falls: Biodiversity Bonus?

The discovery of hydrothermal vents in 1977, and the subsequent discovery of cold seeps in 1984, had a profound effect on the way marine biologists view the deep sea. Up to that point the paradigm was that the deep sea was a heterotrophic system ultimately reliant on surface primary production for its

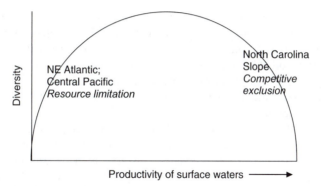

Figure 8 Species diversity as a function of productivity with suggested geographic locations. Modified from ref. 43.

organic energy. Vents and seeps were to prove to be the first major ecosystems independent of sunlight where primary production was carried out *in situ* by endosymbiotic or environmental chemosynthetic bacteria using reduced chemicals such as H_2S and CH_4 as an energy source.[11]

Although vents supported a spectacular biomass, it was apparent from the early days of exploration that actual biodiversity at vents is low,[48] especially when compared to the biodiversity of the continental slope. To date some 470 species are known from vents,[49,50] more than 200 from seeps[51] and a surprisingly high number (407) of species from whale falls.[50] However, more importantly, the species making up the biodiversity at these chemosynthetically driven environments were nearly all new to science. In addition, many new families had to be erected to accommodate these new species. Similarities in genera and families occur across vents, seeps and whale falls but there are significant differences at species level and with species found in the "typical" deep sea.

As a result, vents and seeps contribute locally to increased diversity of the deep sea. Their areal extent is often small and they grade rapidly into the local "normal" deep-sea environments. This is particularly noticeable for cold seeps found mainly along the continental margins.[52]

5 Patterns of Biodiversity in the Deep Sea: Pelagos

Although the volume of the open water deep sea is orders of magnitude greater than the area of the benthos, there is a significant reduction in diversity.[53] Herring suggests that this is partly a function of no phyla being solely pelagic and partly a lack of heterogeneity in the water column. In the benthos (see above) there is considerable heterogeneity at various scales but within the water column, especially below the thermocline, barriers are very limited. Water masses rarely have sharp boundaries and the differences in the physico-chemical variables are very small. Below *c.* 3000 m the pelagos of the different ocean basins may be separated. Such separation is generally required for allopatric speciation and the lack of such separation reduces overall species diversity. However, there is enough difference for the establishment of biogeographical provinces.[53] Global species numbers in the deep pelagic[53] are 2200 species of copepod, 115 chaetognaths, 187 ostracods, 87 euphausids and a few hundred gelatinous zooplankton (medusae and ctenophores). The last group is still expanding as submersible techniques allow collection of these, often very large, delicate species.

From the data available there is evidence for mid-depth increase in species diversity in ostracods[54] together with a maximal biodiversity at 18°N. Such a mid-latitude peak of diversity is observed also in fish, decapods and euphausids and may reflect the overlapping of two faunal provinces.[53] Lastly, pelagic biodiversity is reduced in areas of oxygen deficiency such as those beneath upwelling zones in the Indian and Pacific Oceans.

6   Patterns of Biodiversity in the Deep Sea: Fish

As with benthos and pelagic invertebrates, we are nowhere near an asymptote in understanding fish diversity in the deep sea. 1280 species of fish live below 200 m, of which *c.* 1000 may be pelagic,[53,55] although new descriptions continue on a regular basis[56] with no evidence that the discovery rate is dropping off. Merrett and Haedrich[56] note that it is not cryptic species of fish that are being described but large sharks and rays, never before caught.

The apparent richness of deep-sea fish diversity has led to much discussion on its ultimate cause. Suggestions include different periods of invasion from shallow water, allopatric speciation associated with anoxic events[32,57] and vicariance.[58] This aspect is beyond the scope of this review and the reader is referred to Merrett and Haedrich[56] for a thorough treatment.

Of importance to potential exploitation is the vertical distribution of fish species. To date (2006) deep-sea commercial trawling can occur down to *c.* 2000 m. Merrett and Haedrich review the data on vertical zonation of species within different oceans. They conclude that "almost all" species have a remarkably limited vertical range. For example, in the NE Atlantic nearly 75% of species are found in the top 2000 m with the remainder below 3000 m, except for five species that have a broad vertical range (Figure 9A). This pattern is less obvious off West Africa or in the NE Pacific (Figure 9B). Off West Africa the greatest diversity is at slope depths whereas in the Porcupine Seabight in the NE Atlantic there is a mid-slope diversity peak and a smaller peak at 4200 m.[56]

Lastly, deep-sea fish have *inter alia* two important attributes that may have population consequences from fishing. The first feature is the "bigger-deeper" trend. As a broad rule, although not exclusively, within a species the mean size increases with depth but the abundance decreases. This has potentially serious implications as fisheries extend into very deep water. This appears to be particularly significant beyond 3000 m. The second feature is that many deep-sea fish species, especially the larger and potentially commercially-important ones, have long lives and late maturity and thus low replacement rates under exploited conditions.[56]

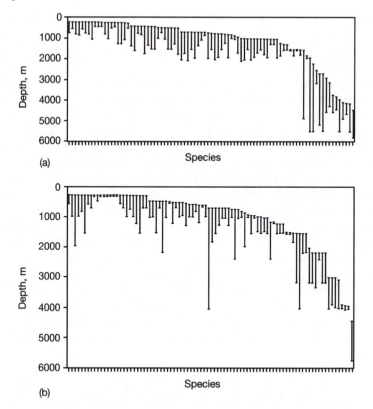

Figure 9   Vertical range of fish (**a**) in the Porcupine Seabight and Abyssal Plain and (**b**)
off West Africa at 8 to 27°, demonstrating that very few fish have a wide depth
range.[56]

7   Is Biodiversity in the Deep Sea Under Threat?

It is difficult to assess the threat to deep-sea biodiversity when the patterns of
biodiversity are still unclear. However, the deep sea is not immune to anthropogenic impact. Two major aspects may have an impact on deep sea biodiversity: disposal and exploitation. The most recent thorough and detailed review
of this aspect is by Thiel.[59] By contrast, Tyler[60] compares the effects of
anthropogenic impact with natural impact at the deep-sea floor. However,
with the limited data available it is difficult to quantify the anthropogenic
activity on biodiversity *sensu strictus* at the deep-sea floor.

7.1   Disposal

Waste has been discarded over the sides of ships since man first sailed the
ocean. In many respects it has been a minor disposal although recent regulations make over-the-side disposal illegal on most western ships. The most

common wastes are oil drums and clinker from the boilers of steam ships, although both have a very little impact on the deep-sea floor except in deep water off major ports where clinker can form relatively thick layers[61] (Tyler, personal observation).

More significant wastes are those of industrial origin such as sewage, dredge spoil and pharmaceuticals that are disposed of in bulk. Certainly waste spoil disposal at the Deep-Water Dumpsite 106 off New York had a profound effect on isotopic composition of the seabed megafauna[62] and increased abundance but not species diversity in the macrofauna. Low-level radioactive waste was deposited at various sites in the deep ocean from the 1950s to its cessation in the 1980s but there has been little monitoring of its impact on biodiversity. What little information there is suggests that the drums containing the waste provided a rare hard substratum in a huge sedimentary area and were occasionally colonised by anemones. The sinking of ships over the abyssal plains would provide a hard, albeit temporary, substratum surrounded by an endless sedimentary environment, thus increasing heterogeneity. Ballard[63] records the presence of gorgonians on the chandeliers of *Titanic* and a soft coral on the stem post, both having settled on a substratum that would not have been expected in the deep NW Atlantic. Thus one could argue that alpha diversity had increased as a result of drum disposal and ships sinking, both increasing the heterogeneity of the local environment!

Of considerable potential future significance is the disposal of carbon dioxide. At certain deep-sea temperatures and pressures CO_2 forms hydrates in liquid or solid form.[64] Early experimentation on the disposal of CO_2 into the deep sea has shown that it will effect an immediate local reduction on biodiversity[65] and, if scaled up to industrial levels, would have a profound detrimental regional impact.

CO_2 disposal also lowers the local pH (ocean acidification) and would thus have a potentially damaging effect on particularly corals, shelled invertebrates and other calcareous organisms in the water column. Predictions are that the calcium carbon compensation depth (where carbonate goes into solution) will move nearer the surface, potentially having a profound effect on biodiversity of calcareous organisms, such as foraminifera, in the water column.

On a much longer time scale there is concern that global warming may affect deep water mass formation and circulation. Most of the water covering the deep-ocean basins originates in the Norwegian Sea.[10] North Atlantic surface water flows into the Norwegian Sea where, in winter, it is cooled by heat loss and mixing with East Greenland Water to form deep water that sinks and eventually flows southward across the Scotland-Iceland-Greenland Ridge into the NW Atlantic where it forms North Atlantic Deep Water[10] that spreads throughout the world ocean, ensuring deep-sea sediments are overlain by oxygenated water. If global warming causes the cessation of deep-water formation in the Norwegian Sea, there will be no deep-water flow and over time metabolic processes at the deep-sea floor will consume oxygen and eventually the deep ocean basins will become anoxic. This has already happened in the Black Sea, although not as a result of global warming, but of greatly reduced circulation and limited inflow of water from the Mediterranean.

7.2   Exploitation

The deep sea contains many potentially commercial biological and mineral resources. Fishing is extending into much deeper water. There has been considerable concern about the effect of fishing on deep-water corals, especially *Lophelia pertusa*, where the coral is broken up by dragging metal over it before a trawl is used. Corals have high local biodiversity and fishing is an ongoing threat, not yet rigorously quantified.[66] Already the Darwin Mounds area of the NE Atlantic has been declared a marine protected area, where no fishing is allowed. Destruction of this site was almost total (Figure 10).

Fishermen are also responsible for the almost total decimation of several species of fish.[67] The removal of these species close to the top of the food chain may have serious consequences for those lower down. Loss of such a resource is occurring before we know anything about the biology of these fish. If they are long-lived and slow recruiting, as expected, recovery may be years away, if ever.

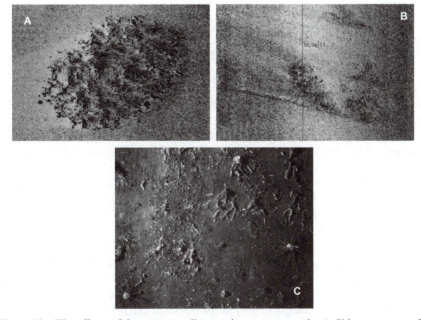

Figure 10   The effects of deep-sea trawling on deep-water corals. **A**. Sidescan sonar of a normal deep-water coral mound.; **B**. Sidescan sonar of heavily trawl-impacted coral, with most of the coral destroyed; **C**. deep-water photograph of the heavily trawl-impacted Darwin Mounds area of the NE Atlantic (*c*. 1000 m depth). © Brian Bett, National Oceanography Centre, Southampton.

Oil exploration is extending into deeper and deeper water and oil companies are particularly sensitive to disturbance of the deep-sea populations. Ironically, some of the deep-water oil production platforms have been colonised by the deep water coral *Lophelia pertusa*. Polyaromatic hydrocarbons are known to have an effect on the immune system of bivalves near oil rigs[68] that makes the bivalves susceptible to parasitic invasion and reproductive senescence. Although the effects of oil production have been tested autecologically there is little evidence for their effect on biodiversity.

Other mineral exploitation concerns manganese nodules and sulfide mineral deposits. Both now come under the jurisdiction of the International Seabed Authority of the United Nations. Sulfide mineral deposits are found mainly associated with hydrothermal vents and there has been little interest in exploiting them.

Of greater interest have been manganese nodules, found on parts of the abyssal plain, as they contain strategic minerals.[69] Manganese nodules can cover 20 to 40% of the abyssal plain (see ref. 70 for photographs), and increase the heterogeneity of the environment. Manganese nodules have their "own" fauna consisting mainly of foraminifera, often stratified on the nodule or in crevices and not usually found in surrounding sediment.[71,72] The impact of nodule mining in deep-sea sediments can be drastic. In pilot experiments with small collectors the sediment and nodules are removed to several cm depth, whilst uncollected sediment is pushed to one side as a levee.[59] Nodules are retained whilst waste sediment is discharged into the lower water column and may be transported up to 10 km before settling. The outcome is that the fauna in the path of the collector is destroyed, fauna nearby is smothered and alpha diversity would be reduced. However, scavengers might be attracted to the dead organisms, so that for a period alpha diversity might increase before decreasing sharply. Opportunistic species might colonise the collector tracks supplementing local biodiversity.[73]

Genuine experimental work on the effects of human impact on the deep-sea bed are very limited. In the 1970s and 1980s there were several experiments that attempted to look at recolonisation of defaunated sediment. Defaunated sediments in recolonisation trays were deployed at 1760 m depth off New England.[74] Samples were taken at 2 and 26 months. The colonising species rarely represented the background species diversity and were an order of magnitude fewer in number. Ironically, the most common species in the 26-month tray was an ectoparasitic isopod! The next four most common species were predators. These data suggested that recolonisation was extremely slow in the deep sea and disturbance caused a severe reduction in deep-sea fauna that resulted in a long-lasting source of spatial heterogeneity.[74] In a subsequent series of similar experiments deployed for 6 months at 2160 m depth in the Bay of Biscay, intensive recolonisation of defaunated deep-sea sediment was observed when compared to neritic sediment with high organic matter.[75] In contrast to Grassle, Desbruyères found a density of fauna five times the background community, although the diversity was greater in the background community (see Table in ref. 75). In a later study conducted at 2160 and 4100 m in the Bay of Biscay, the recolonisation rate for microflora, meiofauna and macrofauna was "generally slow".[76] In long-term deployments to determine colonisation at 1800 and 3600 m depth off New England, neither the diversity nor the density of individuals in experimental trays reached that of natural sediment after 5 years.[77] However, patches of organic material such as rotting seaweed increased recolonisation, particularly by opportunistic species. The recolonisation trays deployed for the longest period of time did have the highest diversity, suggesting that the recolonising communities were converging with the natural background.[77] These early small-scale experiments, however, did not reflect the true disturbance that would be seen if mining was to take place at the seabed. This experimental approach was continued by deploying artificial enrichments and depressions at the deep-sea floor.[78] The resulting data showed some profound differences in colonisation of both the enriched trays and the depressions and it was concluded that the "patch mosaic" was a critically important part of maintaining biodiversity in the deep sea. However, a more recent review of the experimental and comparative evidence[79] showed little or no change in diversity from disturbance or increased food input.

In the 1980s there was a move to longer-term studies of potential impacts on deep-sea biodiversity. Before considering these, it is necessary to determine exactly "what is the baseline?" At present, a baseline study would be what we find today at the deep-sea floor. However, using the example of the 4800-m-deep Porcupine Abyssal Plain in the NE Atlantic, a baseline study today would yield a very different baseline from that observed in the 1980s. The so-called "*Amperima* event"[80] has changed the baseline dramatically over less than a decade. This event has changed the dominant macro- and megabenthic species on the Porcupine Abyssal Plain such that it would be difficult to recognise the community from the 1980s (see Figure 11 and Table 1).

In the 1970s two major programmes addressed impacts on the deep sea: DOMES (Deep-ocean Mining Environmental Study) and MESEDA (Metalliferous Sediments Atlantis II Deep). These used tested oceanographic methodologies to determine the effect of metalliferous mining on the deep sea. Both resulted in huge datasets but were unable to support predictions of impacts.[59] This disadvantage was addressed in the DISCOL project.[59,81]

In the long-term study (DISturbance-COLonisation Study), Thiel and co-workers used a harrow to plough up *c.* 11 km^2 of seabed 600 km south of the Galapagos Islands (88°W 9°S) at a depth of 4150 m.[59,82] Before ploughing, a baseline seabed survey was undertaken, followed by subsequent surveys at 6 months, 3 and 7 years after the ploughing.[59] After 7 years the plough tracks were still obvious.

Faunal composition of all size classes was permanently altered by the ploughing and the manganese nodules were ploughed into the sediment. Megafauna recovered slowly, with a strong increase in numbers at 3 years; at 7 years even some of the less mobile species had returned.[59,83]

Macrofauna were very variable in their recovery. Initial recovery was rapid but slowed with time from the disturbance.[59] Peracarids recovered rapidly but polychaetes were still severely affected after 3 years. Species diversity calculated

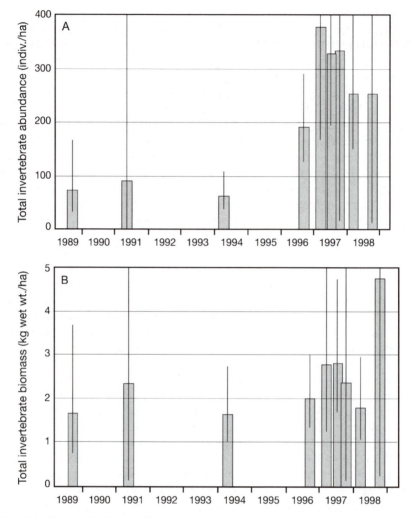

Figure 11 Temporal changes (the *Amperima* event) at 4800 m on the Porcupine Abyssal Plain in the NE Atlantic. **A**. Abundance; **B**. Biomass. There was also a marked change in the dominant fauna (see Table 1).[80]

from 78 polychaete species in disturbed samples was significantly different from that in the undisturbed area. This is believed to be a function of the artificial heterogeneity introduced into the area by the ploughing. Bivalves also displayed an erratic recovery pattern. Initially, abundance increased rapidly but after 3 years started to decline again.[59] Of interest is that the meiofauna abundance increased over pre-ploughing levels but diversity reduced as the manganese nodules were buried, taking with them the meiofaunal communities unique to them (see above). Thiel[59] reported on more recent larger scale projects but opined "that the results from these experiments are so far limited".

Table 1 Changes in the faunal composition with time at 4800 m depth on the Porcupine Abyssal Plain, NE Atlantic.[79]

Station number	52701#61	52916#1	53205#14	13078#47	13200#95	13370#8	54904#2
Frames analysed							
Larger fauna	1165	1712	806	1860	1146	1605	1790
Ophiuroids	40	40	20	185	1146	162	172
Numbers observed							
Ophiuroids	13	40	5	448	9014	1487	569
Amperima rosea	0	0	0	692	690	476	614
Other holothurians	13	37	5	23	38	315	50
Other megabenthos[a]	5	8	1	2	0	6	8
Abundance (/ha)							
Ophiuroids	5370	16,520	4131	40,007	109,820	151,641	54,652
Amperima rosea	0	0	0	6695	8286	4782	6552
Other holothurians	55	107	31	61	164	972	138
Other megabenthos[a]	21	23	6	5	0	19	22
Total megabenthos[a]	3812	11,624	2911	34,597	82,025	111,278	44,737

One area of potential effect on deep-sea biodiversity is through the actions of scientists themselves. Although scientists are unlikely to have a major impact sampling the abyssal plains, there may be an impact on local diversity from collecting at small-scale environments such as vents or coral mounds. The possibility that scientists were having an impact at vent sites was first raised in the late 1980s.[84] This theme was expanded[85] but later refuted,[86] the latter authors proposing that scientists were careful during collections and that vents could also be wiped out in minutes by lava eruptions at the seabed as has been witnessed on a number of occasions (most recently June 2006 at 9°N on the East Pacific Rise). However, there is now a working code of practice for scientists working at vent and seeps set up to ensure there is no diminution in the diversity at these environments. Another potential preventive measure is the establishment of "Marine Protected Area" or Marine Parks. This has already occurred for part of the vent systems in Portuguese waters south of the Azores and for a part of the NE Pacific vent biogeographic area.

8 Conclusions

Our understanding of deep-sea biodiversity is best summed up in the philosophy of the Census of Marine Life (CoML) field programmes ChEss, Cedamar, Mar Eco and Comarge (see www.coml.org for a portal to these programmes). These programmes are addressing the known, unknown and the unknowable. The unknown can be discovered and quantified but the unknowable cannot be assessed. It is the unknowable that remains the black hole in understanding deep-sea biodiversity.

Recent estimates of deep-sea biodiversity have ranged from 5×10^5 to 10^8, all based on samples from an insignificant area when compared to the area of the deep-sea floor. There is no doubt that, as we understand heterogeneity over a variety of scales in the deep sea, this number will be refined and, hopefully, become more accurate. Molecular methodologies are being used to synonymise species collected and described by different institutions but also to identify cryptic species in particularly speciose taxa such as the vent and seep bivalve genus *Bathymodiolus*.[87]

Until we know what the diversity is in even a particular area of the deep sea it is difficult to determine whether it is under threat. The most obvious examples of species under threat involve fishing activity. Deep-water corals have a high associated diversity but their destruction by fishing gear is becoming very evident. Although they occupy a large area of the upper slope, areas such as the Darwin Mounds in the NE Atlantic have been declared Marine Protected Areas in an attempt to preserve the present biodiversity. Understanding the autecology of individual species helps us understand the stability of the entire community and predict which species, if any, are most prone to anthropogenic impact because of their life history tactics. The evidence to date suggests there is no "typical" deep-sea life history, making the threat of impact on biodiversity even more difficult to predict. This becomes particularly important as human exploitation for mineral resources moves into deeper and deeper water.

At present all we can say is that the deep sea, with few exceptions, has a remarkable resilience to any insult imposed on it. Even for those areas where there appears to be a real effect we cannot tell what the long-term effect will be, whether recovery or loss, because of the nature of deep-sea organisms.

Acknowledgements

I wish to thank Mike Rex (University of Massachusetts, Boston) and John Lambshead (Natural History Museum, London) for many interesting discussions on deep-sea biodiversity and Hjalmar Thiel for similar discussions of anthropogenic impacts. Mike Rex was kind enough to make many useful comments on the manuscript. Brian Bett (National Oceanography Centre, Southampton) kindly provided Figure 10.

Glossary of Technical terms

Abyssal: between 3000 and 6000 m deep; the abyssal plain
Allopatric: not occurring together
Alpha diversity: biodiversity within one sample or station
Bathyal: between 200 and 3000 m deep; the continental slope
Benthos: animals living at the seabed
Macrofauna: animals retained on a 0.42-mm mesh sieve but not visible in bottom photographs.
Megafauna: the animals visible in bottom photographs.
Meiofauna: animals that pass through a 0.42-mm mesh sieve.

Oligotrophic gyres: the centres of the main ocean where production is low

Pelagos/pelagic: animals living in the water column

Thermohaline circulation: water mass movement driven by temperature and salinity differences

Upwelling: where nutrient-rich water from depth (usually < 200 m) rises to the surface and increases surface primary production

Vicariance: the geographic separation of a species over geological time resulting in two closely related species, one being the geographic counterpart of the other.

References

1. P. A. Tyler, *Ocean. Mar. Biol. Ann. Rev.*, 1980, **18**, 125–153.
2. E. Forbes, "Report to the 13th meeting of the British Association for the Advancement of Science", 1943, **1844**, 30–193.
3. E. L. Mills in "The Sea", G. T. Rowe (ed.), Wiley Interscience, New York, 1983, **8**, 1–79.
4. C. W. Thomson, "Depths of the Sea", McMillan, London, 1874.
5. P. A. Tyler, *Ocean. Mar. Biol. Ann. Rev.*, 1995, **33**, 221–224.
6. C. D. Hollister and I. N. McCave, *Nature*, 1984, **309**, 220–225.
7. D. Thistle and G. D. F. Wilson, *Deep Sea Res.*, 1987, **34**, 73–87.
8. A. D. Rogers, *Adv. Mar. Biol.*, 1994, **30**, 307–350.
9. A. Genin, P. K. Dayton, P. F. Lonsdale and F. N. Speiss, *Nature*, 1986, **322**, 59–61.
10. J. D. Gage and P. A. Tyler, "Deep-sea Biology: A Natural History of Organisms at the Deep-sea Floor", Cambridge University Press, Cambridge, 1991.
11. C. L. Van Dover, "The Ecology of Deep-sea Hydrothermal Vents", Princeton University Press, Princeton, New Jersey, 2000.
12. G. T. Rowe and N. Staresinic, *Ambio Special Report*, 1979, **6**, 19–23.
13. L. A. Levin, *Ocean. Mar. Biol. Ann. Rev.*, 2003, **41**, 1–45.
14. R. R. Hessler and H. L. Sanders, *Deep Sea Res.*, 1967, **14**, 65–78.
15. H. L. Sanders, R. R. Hessler and G. R. Hampson, *Deep Sea Res.*, 1965, **12**, 845–867.
16. H. L. Sanders, *Am. Nat.*, 1968, **102**, 243–282.
17. J. H. Connell, *Science*, 1978, **199**, 1302–1309.
18. J. F. Grassle and N. J. Maciolek, *Am. Nat.*, 1992, **139**, 313–341.
19. P. A. Jumars, *Mar. Biol.*, 1975, **30**, 253–266.
20. R. J. Etter, M. A. Rex, M. R. Chase and J. M. Quattro, *Evolution*, 2005, **59**, 1479–1491.
21. R. M. May, *Nature*, 1992, **357**, 278–279.
22. P. J. D. Lambshead and G. Boucher, *J. Biogeog.*, 2003, **30**, 475–485.
23. C. T. Stuart, M. A. Rex and R. J. Etter in "Ecosystems of the Deep Sea", P. A. Tyler (ed.), Elsevier, Amsterdam, 2003, **28**, 295–311.
24. M. A. Rex, *Ann. Rev. Ecol. Syst.*, 1981, **12**, 331–353.

25. M. A. Rex in "The Sea", G. Rowe (ed.), Wiley, New York, 1983, **8**, 453–472.
26. M. R. Chase, R. J. Etter, M. A. Rex and J. M. Quattro, *Biotechniques*, 1998, **24**, 243–245.
27. J. D. Zardus, R. J. Etter, M. R. Chase, M. A. Rex and E. E. Boyle, *Mol. Ecol.*, 2006, **15**, 639–651.
28. K. L. Howell, D. S. M. Billett and P. A. Tyler, *Deep Sea Res.*, 2002, **49**, 1901–1920.
29. M. Huston, *Am. Nat.*, 1979, **113**, 81–101.
30. R. F. L. Self and P. A. Jumars, *J. Mar. Res.*, 1988, **46**, 119–143.
31. M. A. Rex, C. R. McClain, N. A. Johnson, R. J. Etter, J. A. Allen, P. Bouchet and A. Waren, *Am. Nat.*, 2005, **165**, 163–178.
32. A. D. Rogers, *Deep Sea Res. II*, 2000, **47**, 119–148.
33. M. V. Angel in "Marine Biodiversity: Patterns and Processes", Cambridge University Press, Cambridge, 1997, 35–68.
34. M. A. Rex, C. T. Stuart, R. R. Hessler, J. A. Allen, H. L. Sanders and G. D. F. Wilson, *Nature*, 1993, **365**, 636–639.
35. S. J. Culver and M. A. Buzas, *Deep Sea Res.*, 2000, **47**, 259–275.
36. P. J. D. Lambshead, J. Tietjen, T. Ferrero and P. Jensen, *Mar. Ecol. Prog. Ser.*, 2000, **194**, 159–167.
37. M. A. Rex, C. T. Stuart and R. J. Etter, *Mar. Ecol. Prog. Ser.*, 2001, **210**, 297–298.
38. T. Brey, M. Klages, C. Dahm, M. Gorny, J. Gutt, S. Hahn, M. Stiller, W. E. Arntz, J. -W. Wagele and A. Zimmermann, *Nature*, 1994, **368**, 297.
39. J. A. Blake and B. E. Narayanaswamy, *Deep Sea Res. II*, 2004, **51**, 1791–1815.
40. J. D. Gage, *Deep Sea Res. II*, 2004, **51**, 1689–1708.
41. M. A. Rex, R. J. Etter and C. T. Stuart, "Large-scale Patterns of Species Diversity in the Deep-sea Benthos", Cambridge University Press, Cambridge, 1997.
42. R. B. Waide, M. R. Willig, C. F. Steiner, G. Mittlebach, L. Gough, S. I. Dodson, J. P. Juday and R. Parmenter, *Am. Nat.*, 1999, **30**, 257–300.
43. L. Levin, R. J. Etter, M. A. Rex, A. Gooday, C. R. Smith, J. Pineda, C. T. Stuart, R. Hessler and D. L. Pawson, *Ann. Rev. Ecol. Syst.*, 2001, **32**, 51–93.
44. C. R. Smith, W. Berelson, D. J. Demaster, F. C. Dobbs, D. Hammond, D. J. Hoover, R. H. Pope and M. Stephens, *Deep-Sea Res. II*, 1997, **44**, 2295–2317.
45. S. Honjo, J. Dymond, R. Collier and S. J. Manganini, *Deep Sea Research II*, 1995, **42**, 831–870.
46. P. J. D. Lambshead, C. J. Brown, T. Ferrero, N. J. Mitchell, C. R. Smith, L. E. Hawkins and J. Tietjen, *Mar. Ecol. Prog. Ser.*, 2002, **236**, 129–135.
47. A. Glover, C. R. Smith, G. J. L. Paterson, G. D. F. Wilson, L. Hawkins and M. Sheader, *Mar. Ecol. Prog. Ser.*, 2002, **240**, 157–170.
48. J. F. Grassle, *Adv. Mar. Biol.*, 1986, **23**, 301–362.
49. V. Tunnicliffe, A. G. McArthur and D. McHugh, *Adv. Mar. Biol.*, 1998, **34**, 353–442.

50. C. R. Smith and A. R. Baco, *Ocean. Mar. Biol. Ann. Rev.*, 2003, **41**, 311–354.
51. M. Sibuet and K. Olu, *Deep Sea Res. II*, 1998, **45**, 517–567.
52. V. Tunnicliffe, S. K. Juniper and M. Sibuet in "Ecosystems of the World: Ecosytems of the Deep Sea", Elsevier, Amsterdam, 2003, **28**, 81–110.
53. P. Herring, "The Biology of the Deep Ocean", Oxford University Press, Oxford, 2002.
54. M. V. Angel in "Oceanography: an Illustrated Guide", Manson Publishing, London, 1996, 228–243.
55. D. M. Cohen, *Proc. Cal. Acad. Sci.*, 1970, **38**, 371–379.
56. N. R. Merrett and R. L. Haedrich, "Deep-sea Demersal Fish and Fisheries", Chapman and Hall, London, 1997.
57. B. N. White, *Biol. Ocean.*, 1987, **5**, 243–259.
58. G. J. Howes, *J. Biogeog.*, 1991, **18**, 595–622.
59. H. Thiel in "Ecosystems of the World: Ecosystems of the Deep Sea", P. A. Tyler (ed.), Elsevier, Amsterdam, 2003, 427–471.
60. P. A. Tyler, *Env. Cons.*, 2003, **30**, 26–39.
61. R. B. Kidd and Q. J. Huggett, *Oceanol. Acta*, 1981, **4**, 99–104.
62. C. L. Van Dover, J. F. Grassle, B. Fry, R. H. Garit and V. R. Starczak, *Nature*, 1992, **360**, 153–156.
63. R. D. Ballard, "The Discovery of the Titanic", Warner Books, New York, 1987.
64. P. G. Brewer, G. Friedrich, E. T. Peltzer and F. M. Orr, *Science*, 1999, **284**, 943–945.
65. D. Thistle, K. R. Carman, L. Sedlacek, P. G. Brewer, J. W. Fleeger and J. P. Barry, *Mar. Ecol. Prog. Ser.*, 2005, **289**, 1–4.
66. A. D. Rogers, *Int. Rev. Hydrobiol.*, 1999, **84**, 315–406.
67. J. A. Devine, K. D. Baker and R. L. Haedrich, *Nature*, 2006, **439**, 29.
68. E. N. Powell, R. D. Barber, M. C. Kennicutt II and S. E. Ford, *Deep Sea Res.*, 1999, **46**, 2053–2078.
69. G. R. Heath in "The Environment of the Deep Sea", Prentice-Hall, Englewood Cliffs, New Jersey, 1982, 105–153.
70. B. Heezen and C. D. Hollister, "The Face of the Deep", Oxford University Press, New York, 1971.
71. L. S. Mullineaux, *Deep Sea Res.*, 1987, **34**, 165–184.
72. H. Thiel, G. Schriever, C. Bussau and C. Borowski, *Deep Sea Res.*, 1993, **40**, 419–423.
73. P. A. Jumars, *Mar. Min.*, 1981, **3**, 213–229.
74. J. F. Grassle, *Nature*, 1977, **265**, 618–619.
75. D. Desbruyères, J. Y. Bervas and A. Khripounoff, *Oceanol. Acta*, 1980, **3**, 285–291.
76. D. Desbruyères, J. W. Deming, A. Dinet and A. Khripounoff in "Peuplements Profonds du Golfe de Gascoigne", L. Laubier and C. Monniot (eds), IFREMER, 1985, 193–208.
77. J. F. Grassle and L. S. Morse-Porteous, *Deep Sea Res.*, 1987, **34**, 1911–1950.

78. P. Snelgrove, J. F. Grassle and R. F. Petrecca, *J. Mar. Res.*, 1994, **52**, 345–369.
79. P. V. R. Snelgrove and C. R. Smith, *Ocean. Mar. Biol. Ann. Rev.*, 2002, **40**, 311–342.
80. D. S. M. Billett, B. Bett, A. L. Rice, M. H. Thurston, J. Galéron, M. Sibuet and G. Wolff, *Prog. Oceanogr.*, 2001, **50**, 325–348.
81. H. Thiel and G. Schriever, *Ambio*, 1990, **19**, 245–250.
82. H. Thiel, *Mar. Min.*, 1991, **10**, 369–386.
83. H. Bluhm, G. Schriever and H. Thiel, *Mar. Geores. Geotech.*, 1995, **13**, 393–416.
84. V. Tunnicliffe, *J. Geophys. Res.*, 1990, **95**, 12961–12966.
85. M. Johnson, *Nature*, 2005, **433**, 105.
86. P. A. Tyler, C. R. German and V. Tunnicliffe, *Nature*, 2005, **434**, 18.
87. J. Jones, Y. -J. Won, P. A. Y. Maas, P. J. Smith, R. A. Lutz and R. C. Vrijenhoek, *Mar. Biol.*, 2005, DOI 10.1007/s00227-005-0115-1.

Threatened Habitats: Marginal Vegetation in Upland Areas

ALISON HESTER AND ROB BROOKER

1 Introduction

What is a marginal habitat? It is a widely used term but is seldom specifically defined. Examples include road verges and field "islands", field margins, ditches, stone walls and ponds.[1] The term has also been used, for example, to describe isolated populations of *Fraxinus excelsior* on islands at the northern limit of distribution of this species.[2] Marginal habitats are, therefore, marginal either in terms of their location relative to dominant habitats (*e.g.* on the edge of other, larger habitats) or marginal in terms of low coverage (marginal for persistence). Summarising the most common uses of the term, our working definition for this chapter is as follows: marginal habitats are isolated pockets of a vegetation type which are: (a) on the edge of, or surrounded by other, more widespread habitats; and/or (b) on the edge of a species range, *i.e.* a habitat that is marginal for a particular species.[3–6]

  Given the above definition, what is the importance of marginal habitats for biodiversity and why are such habitats often threatened? Marginal habitats are generally valued for their biodiversity as they combine unique assemblages of plant and animal species due to their location at the edges of other, often more widespread, habitats and/or species ranges.[7,8] Moreover, as they often span the "transition" between habitats with very different structures, such as high forest and open ground (see Figure 1), they may consist of mosaic vegetation at different heights and densities, thus providing niches exploited by plants and animals not present in adjacent habitats, for example blue throat (*Luscinia svecica*) and black grouse (*Tetrao tetrix*) in altitudinal tree-line zones.[9] The biodiversity resource of marginal habitats is therefore more than just the sum of the two bordering habitats. The very features indicated in the definitions above make marginal habitats vulnerable to environmental and other changes. Small

Issues in Environmental Science and Technology, No. 25
Biodiversity Under Threat
Edited by RE Hester and RM Harrison
© The Royal Society of Chemistry, 2007

Figure 1   Marginal tree-line habitat; structurally diverse vegetation dominated by *Pinus sylvestris*. Scotland, UK. Photograph: Alison Hester.

changes in climate, for example, can quickly render local conditions unsuitable for species at the edges of their range, or can create more favourable conditions for superior competitors to increase their range at the expense of the marginal habitat. Narrow belts or small, isolated pockets of suitable habitat are also vulnerable to a wide range of pressures because the component species may have nowhere else to "go" if the local conditions become unsuitable, be that through climatic or land-use change. Furthermore, fragmentation of habitats, which may result from a wide variety of environmental change drivers (*e.g.* tourism, land-use change and climate change) and which can have particularly severe negative consequences for long-term ecosystem persistence, is likely to strongly affect ecosystems with the type of linear spatial arrangement found in many marginal habitats.

2   Case Studies

To aid our discussion of the issues relating to marginal habitats and the threats to their biodiversity, we have chosen two upland case studies which we will use, along with other examples, to illustrate the different points made through this chapter: sub-arctic willow (*Salix* spp.) scrub and mountain pine (*Pinus mugo*) scrub. Both of these habitats occupy relatively narrow belts between the upper climatic limits of forest growth and the low alpine heaths and grasslands, making them relatively vulnerable to climatic change as well as land-use pressures. The exact combination of climatic conditions facilitates the domi-nance of species that would otherwise be either out-competed (in more favour-able environmental conditions) or unable to grow because of severely adverse environmental conditions. The upper altitudinal limits of both habitats are often (but not always) climatically controlled, whereas the lower limits are generally fixed by competitive displacement by other species.[10,11] Such processes also occur at the latitudinal limits of tree growth, but it is the steep topography and associated steep climatic gradients that create narrow bands of scrub habitat in alpine and montane environments. Sub-arctic willow scrub is rare but mountain pine scrub is not, giving an interesting contrast for comparison. It is worth noting here that, although both are termed "habitats" in the context of this chapter, the former has several key plant species but the latter is dominated by a single vascular plant species, *i.e. Pinus mugo*, which can lead to some confusion when the word "habitat" is still used. The EU Habitats Directive has many other examples of this "blurring" of the terms "habitat" and "species" – we stick with their terminology and use "habitat" for both our examples in this chapter.

Sub-arctic willow (*Salix spp.*) scrub (Figure 2) is defined as a rare habitat, occurring above the altitudinal/latitudinal forest limits in mountain areas of the UK and Scandinavia, dominated by willow species: *Salix lapponum, S. myrsinites, S. arbuscula, S. lanata, S. reticulata, S. glauca, S. myrsinifolia* and *S. phylicifolia*.[12–14] Willows are dioecious, *i.e.* they have separate male and female plants, which makes isolation critically limiting for sexual reproduction, although they readily spread by clonal layering. The habitat is positively associated with areas of moderate–late snow lie and shelter (in the more exposed parts of its range).[13–15] Many of the associated willow species prefer base-rich soil, which also limits their distribution in parts of their range where such rock types are less common.[13,14] Sub-arctic willow scrub is classified as of

Figure 2 Sub-arctic willow scrub, Stoldalen, Norway. Photograph: David Mardon.

high conservation importance across Europe, *i.e.* an Annex I habitat (code: 31.622) in the EU Habitats Directive.[16] Similar types of willow scrub, but without the arctic species (such as *Salix lanata*) are also found (rarely) in the mountain ranges of central and southern Europe. A range of similar sub-alpine willow habitats are also found across North America.

Sub-arctic willow scrub is defined as having relatively high biodiversity value.[17–19] Many willow species are rich hosts for insect herbivores[20,21] and it has been suggested from the limited research to date that their fungal diversity may also be very high.[19] In Scotland, willow scrub communities have been described as supporting one of the highest diversities of vascular plants of all the upland communities.[22] Importantly, the presence of low shrubs in alpine and arctic habitats can act as a protection to other plant species which may thrive under their canopies in these harsh environments,[23,24] although detrimental as well as positive effects have been found for several species studied,[25] primarily due to competition for light.

Although sub-arctic willow scrub is classed as rare over much of its range, in many areas it is not considered threatened; its rarity is simply attributed to a relatively narrow niche between the more widespread montane and forest habitats. However, this is not the case in the UK where a combination of extreme rarity and isolation already threatens the future of this habitat, regardless of any future changes in climate or land use. The limited data available for the UK[26] indicate an overall decline (since pre-1930 records) in the populations of all willow species associated with this habitat, but the lack of good survey data makes it impossible to say with any certainty how the status of this habitat has changed in the UK.[26,27]

Mountain pine (*Pinus mugo*) scrub (Figure 3) is widespread across the high sub-alpine regions (mostly between 1000 and 2500 m altitude) of central and southern Europe from the Swiss/Austrian border (Alps) south-east through Serbia, Romania and Bulgaria and east to the Carpathian mountains, with an isolated population in the central Italian Apennines and "outliers" in the French Alps and Vosges.[28] It is also found at lower altitudes in bogs and frost hollows. Winter snow cover provides important protection for this species.[29] Two sub-species have been recognised (some papers actually describe them as separate species): *P. mugo* ssp. *mugo* in the south and east of its range and ssp. *uncinata* in the north and west (Pyrenees to Poland). The former tends to be small (3–6 m in height) and multi-stemmed, the latter taller growing, often with single stems, but the two species hybridise extensively. The growth form of *P. mugo* ssp. *mugo* is low and shrubby, considered to be an adaptation to deep snow cover and avalanches (*cf. Pinus pumila*[30]). Most *P. mugo* papers do not define the subspecies, so we work only at the species level in this chapter and assume that most are referring to ssp. *mugo*.

The biodiversity value of mountain pine habitat has received limited attention across its range. Examples include records of higher lichen diversity than is found in several other high-altitude habitats (*e.g.* Dolomites, NE Italy[31]) and indications that it also contains relatively high diversity of lower plants in general.[32] Importantly, the value of a habitat cannot be assessed only by the

Figure 3 *Pinus mugo* scrub, Dolomites, Italy. Photograph: Alison Hester.

absolute number of species that it contains. Other aspects of diversity promoted by mountain pine scrub include: (a) the provision of favourable habitat for specific rare species, such as *Rhododendron hirsutum*, which is mainly restricted to these scrub habitats (particularly on calcareous substrates) and (b) its role as a key component of an overall landscape mosaic supporting valued, wider-ranging bird and animal species, such as blue throats (*Luscinia svecica*). It has also been proposed that mountain pine scrub plays an important role in the functioning of the natural environment by protecting the soil and stabilising snow cover. In doing so this habitat also provides important "ecosystem services", such as protecting karstic drinking water catchments by increasing water retention capacity and reducing surface runoff,[33] as well as restricting the release of avalanches and providing suitable conditions for other species.[34]

Pollen analysis indicates that mountain pine scrub was more widespread prior to about 4000 BC in the mountains of Bulgaria[35,36] and the occidental Alps of Italy,[37] with the subsequent decline being attributed to combinations of climatic change and increased human interference (burning and grazing). Across its core range, mountain pine scrub is not considered to be endangered:

although some of the isolated populations are thought to be declining, in other areas its range has expanded, resulting in no strong trends overall.[38]

3   Drivers, Pressures and Threats

Identification of the main threats to the biodiversity of marginal habitats requires a good understanding of the drivers and pressures of change (see other chapters in this book for further, complementary discussion). The concept of separating drivers and pressures of change has developed from the DPSIR framework (see below for definition). Originally developed in the context of social studies, DPSIR is now being utilised to explore environmental issues, for example in the development of state-of-the-environment reports. The European Environment Agency has adopted the DPSIR framework and defines it as a "causal framework for describing the interactions between society and the environment".[39] The letters refer to D – driving forces, P – pressures, S – states, I – impacts, R – responses. In the context of ecological issues it is often difficult to separate drivers and pressures. Indeed, we consider many of the key environmental change processes, such as climate change, land-use change and tourism, to be a complex combination of the two. Therefore we do not attempt to separate them, but simply refer to both as *drivers*. Figure 4 shows how the DPSIR framework can be used to structure the process of integrating biodiversity change management, research and policy development

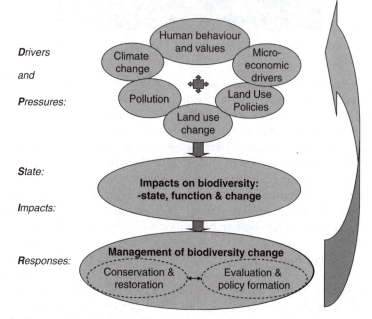

Figure 4   Integrated approach to biodiversity management, research and policy development: a DPSIR framework.

(compare Figure 5 in Chapter 3). This must be a constantly iterative process, as indicated by the right-hand-side arrow. In brief, human behaviour and values drive policy and land management and these, in interaction with environmental drivers such as climate, combine to drive biodiversity change. Research into impacts on biodiversity and options for conservation/restoration is essential to provide underpinning information which then shapes policy change and completes the iterative process.

Because of their isolation and marginality, our two case-study habitats are likely to be highly vulnerable to a range of environmental change drivers. However, although the threat from a given environmental driver such as climate change might at first seem obvious, the interactions of different environmental drivers can also have complex and sometimes unexpected impacts on these communities. Here we explore some of the key drivers, their potential impacts and evidence of complex interactive effects between them. In particular, we concentrate on climate change and land-use change (particularly grazing) as these have been identified as major drivers of biodiversity and ecosystem processes,[40] particularly in mountain ecosystems.[41]

3.1 Climate Change

At a coarse level, the influence of climate on upland scrub ecosystems is clear. Climate regulates tree-lines by determining the point at which increased environmental hazard (damage) and material limitations (*e.g.* limitations in resource acquisition) operate together to make "accretion of biomass untenable".[42] This is why we see initially a loss of the genuine "tree" growth form with increasing altitude/exposure and decreasing temperature, eventually leading to formation of a band of scrubby marginal habitat (although because of the impact of other drivers not every alpine tree-line has this scrubby margin – see Section 3.1.1 below). Therefore it is logical to expect that under climate warming we would see an upward movement of the tree-line and an upward movement of the associated marginal scrubby habitats above the tree-line, *e.g.* our two case-study examples.[43] Certainly, in view of the narrow climatic zone occupied by these marginal scrub habitats, some current areas are likely to become climatically unsuitable for occupation following climate warming. For example, a recent climate envelope modelling study[44] predicted that nearly all of the current range of mountain pine in a selected area in the low Tatras Mountains will become unfavourable, necessitating a future upward shift in species distributions in order for them to survive. In some cases there is already good evidence of an upward altitudinal shift in tree ranges as a response to recent climate change. For example, evidence for sizable shifts in Scandinavian tree-lines has been found,[45] most probably resulting from the combined effect of recent warm winters and summers. However, expansion of the "absolute species limits" (*i.e.* the occurrence of seedlings at higher altitudes) need not necessarily lead to afforestation. Following seedling establishment, limits will still be imposed on tree growth by other important factors, particularly wind[46] as well as ice-crystal abrasion and extremes of temperature, as soon as individuals grow above the more "favourable" climate near the ground.[47,48] In oceanic islands such as the UK the limits of the tree-line occur at higher mean temperatures than would be expected from studies of continental alpine areas. Higher wind speeds and greater cloudiness in oceanic areas are considered to be the main cause of this discrepancy.[49,50]

Another major climate-related driver of change in upland marginal habitats is snow cover. Winter snow cover has been shown to play a strongly protective role against extremely low temperatures and high light intensities (which can inhibit repair processes in the plant) for species such as *P. mugo* and *Salix* spp.[13–14,29] However, late snow-lie in spring has been shown to increase fungal infection rates in mountain pine[51] and can hamper spring photosynthesis.[52] Stem samples from *Salix arctica* in north-east Greenland also indicated significant negative impacts of spring snow-lie on annual radial growth of this species.[53] So the predicted timing, as well as magnitude, of changes in snow cover will be crucial in determining the likely effects of climate change on these montane scrub habitats.

As well as the direct responses of plant species to climate, interactions with the dominant surrounding vegetation also regulate climate change responses in these habitats. Shrubs are predicted to respond more slowly to increases in temperature than faster growing forbs and graminoids. This could potentially result in greater vulnerability of these marginal scrub habitats to competition from surrounding non-woody vegetation,[54] which could work in two ways: there could be encroachment of other plant species into the scrub zone, and/or failure of the scrub to move upslope (even where climatic conditions are suitable) because of increased competition from other "resident" vegetation which has adapted more quickly to temperature increases. Studies of *Pinus sylvestris* distributions in the Swiss Central Alps following warm summers, for example, showed no signs of the expected upslope movement of trees. In this case it was concluded that species interactions stabilised the position of the habitat and that upslope seedling establishment would need disturbance to create suitable germination niches due to the dense sward of the high altitude grassland above the tree-line.[55] However, in contrast, it has been found[45] that certain tree species seem to have tracked climatic conditions more sensitively than predominantly clonally reproducing herb or graminoid species, so the theory does not always hold. The invasibility and adaptability of the surrounding vegetation are key issues in determining the probable impacts of climate warming on these marginal scrub habitats. Resistance of surrounding vegetation to invasion by these scrub species under conditions of climatic change could simply lead to further fragmentation and decline in these habitats, but could alternatively result in "catastrophic shifts" to other vegetation types,[56] requiring active and costly intervention to secure the future of these marginal scrub habitats.

Research on photosynthetic functions of mountain pine and other species during winter and spring, when most climatic stresses occur,[29] indicated a relatively high stability of this species in response to changing conditions, compared to several other woody species. This indicates good adaptability of mountain pine to climatic change. The phenological strategy of different species (*e.g.* in relation to temperature "cues" for bud-burst, flowering, *etc.*) is considered a key determinant of likely responses to climatic change, particularly for species at high elevations.[57] This is of key importance since, under conditions of temperature change, individual species differences within a community could lead to major changes in plant community composition, rather than the "mass movement" of all species in that habitat to a new "range". Therefore, as well as driving potential range shifts, plant responses to climate change will also affect within-habitat composition and biodiversity. Increased temperature and CO_2 concentrations, for example, are generally predicted to increase plant growth in upland habitats, particularly early in the season.[58–60] But these responses will also have a range of knock-on effects on associated species (plant and animal) due to differential phenology, plant-component growth (*e.g.* stem *versus* leaves) under different predicted CO_2/temperature scenarios and differences in chemical responses.[60] Therefore, as well as affecting plant competitive ability *per se*, there will be potential knock-on effects on the whole food chain.

Direct effects of particular climatic components can be complicated by differential interactive effects. This is an important consideration when making predictions about the net effect of different climate change scenarios. For example, additive negative effects of both enhanced UV-B radiation and drought stress on seedling growth of *Salix myrsinifolia* have been found.[61,62] However, significant variability between clones and families, as well as reduced effects in hybrids, also has been observed.[61,62] This illustrates the importance of genetic as well as species variability in determining plant adaptation to changes in climate. Decreased nitrogen concentrations and increased C/N ratios in woody arctic species have been found in response to a rise in temperature,[54] whilst decreased leaf nitrogen content has been found in arctic willows in response to elevated CO_2 concentrations and differential patterns of increased biomass allocation, as compared to temperature effects.[60] It is, therefore, fundamentally important to understand the probable differences in species responses to all components of predicted climate change scenarios.

3.1.1 Climate and Land-use Interactions. Major drivers rarely act in isolation to drive plant community change. The key role of interactions between climate and land-use in determining tree-line and montane scrub dynamics is reflected in the fact that two "types" of tree-lines may be distinguished, *viz.* transitional and abrupt. Abrupt tree-lines are characterised by a sharp edge, considered to be a consequence of grazing pressure. Browsing depletes resources of the highest altitude individuals and increases mortality, resulting in an abrupt tree-line of established trees that have reached above the browsing line in times of lower grazing pressure. A transitional tree-line has no fixed edge and is commonly associated with a lighter level of grazing pressure and a wider band of low scrub, as supported by research in Fennoscandia.[63]

Biotic interaction processes *within* the marginal habitats themselves may also have an important role to play in regulating the rate at which their altitudinal limits contract or expand during climate change under different land-use scenarios. For example, a negative effect of mountain pine cover on the recruitment and growth of spruce and larch has been found,[11] although it was found to have had a beneficial impact on new tree recruits by providing grazing protection. Therefore, in areas of low grazing one might predict reduced expansion of tree-line forest into areas dominated by mountain pine scrub, but in areas with heavier grazing, forest expansion might be more likely to occur. As well as vegetation limitations to range expansion or movement, soils also limit vegetation change. In addition to the obvious requirements of pH, *etc.*, soils beneath tree-line scrub have been found to be between 0.5 and 4°C cooler than in neighbouring grasslands during the growing season, which would limit the influx of some lower-altitude, more productive species.[11]

In view of the complexity of the processes related to range shifting under climate change, and the interactive effects of land use and climate change, it is not surprising that a generalised scrub expansion has been predicted and observed in tundra and tree-line areas,[64,65] whereas decline has been observed in others.[44] Modelling studies of the distribution of alpine vegetation zones in response to land-use and climate change in the Alps have predicted a severe contraction of the alpine, non-forested zone because of range expansion of mountain pine.[64,66] However, such studies have also shown that the low growth rate of mountain pine, its long generation time and limitations on seed dispersal (low seed release height), should result in comparatively slow range expansion.[38] Furthermore, as discussed above, in many areas the force imposed by a changing climate may be counteracted by resistance of the resident alpine vegetation to invasion.[66] Therefore, it is crucial to assess whether the contraction of the lower range limit will occur at a faster rate than expansion of the upper range limit for these marginal habitats, before being able to predict overall net effects of different climate change scenarios in different areas.

3.2   Grazing

Marginal habitats can be particularly vulnerable to the effects of grazing and browsing, as many associated species are at the edge of their range and therefore may already be growing in sub-optimal conditions and unable to compensate for biomass offtake by herbivores. Willows are highly preferred by herbivores, making sub-arctic willow scrub particularly vulnerable to herbivore grazing pressure.[10,27] For example, significant reductions in willow scrub extent with large increases in red deer (*Cervus elaphus*) populations in Colorado[67] and positive responses to removal of livestock grazing in montane willow communities have been found experimentally, particularly in early years. In the UK, where there is a long history of grazing and sub-arctic willow scrub is currently extremely rare, the remaining patches are mostly very small and confined to rock ledges or other inaccessible areas, often containing only one sex of plant, resulting in no seed production.[10,13,19,27] Individual plants growing within reach of grazing animals tend to be heavily browsed, resulting in zero or negative net annual growth and little or no flowering. Although pines are less preferred by herbivores than are willows,[68] mountain pine has still been shown to be negatively impacted by grazing/pasturing activity.[7,69,70] However, predictions of the probable effects of a range of browsing intensities on mountain pine scrub in eastern Switzerland, using a FORET/JABOWA-type forest succession model,[71] indicated that even heavy herbivore pressure is unlikely to threaten survival of this habitat, although it would alter the structure and mortality rates. The "critical value" quoted for mountain pine is a mean browsing intensity of 30%, below which any changes to the structure and dynamics are not significantly different from those predicted under zero browsing. Complete removal of browsing, however, although often beneficial in the short term, can quickly lead to reduced species and structural diversity of woody vegetation.[68,72] This has also been shown to be the case for montane willows,[73] with declining species diversity and reduced stem recruitment after about 10 years of complete grazing exclusion (even though these willows readily produce new shoots by vegetative layering, as do mountain pines). Therefore, as with forest systems, the most desirable scenario for maximising diversity of these marginal montane scrub habitats is a monitored balance of herbivore densities, according to local conditions, taking into account the reproductive and competitive characteristics of the dominant scrub species.[72,74]

Although there is a tendency to assume that *increased* human activity (particularly through livestock grazing) is a major cause of biodiversity change, in many mountain areas it is now the case that significant changes are occurring as a result of land abandonment. Improving transport infrastructure has enabled more cheaply produced goods to be imported into alpine areas from the lowlands. This has reduced the need for self-reliance in alpine agriculture and has contributed to widespread abandonment of traditional land uses and a decrease in populations.[75] Management activities such as alpine livestock pasturing are known to lead to a decrease in altitude of habitats from above the tree-line, due to grazing pressures.[76] In some cases land abandonment has already been shown to lead to an upward shift ("recovery") in the altitudinal limits of marginal upland scrub habitats. For example, in the Majella massif in Italy the removal of traditional summer pasturing since the 1950s has led to an upward shift in mountain pine scrub.[7,70] Similar evidence exists for mountain pine in Bulgaria[69] and Slovakia.[34] However, as indicated before, if the higher altitude vegetation is resistant to invasion, then even if grazing is reduced it does not necessarily mean that the scrub habitat will expand.[77]

Effects of climate change on herbivore-scrub interactions have been relatively little studied and most work has been done on insects (see below and Chapter 2), but we also can make some predictions for large herbivores. Production of phenolics, for example, is known to deter mammalian herbivores[74,78] and increased concentrations of phenolics have been recorded in some willow species in response to enhanced UV-B, although this was limited by drought stress.[61] One possible (and testable) scenario, therefore, is that these plants might become less attractive to mammalian herbivores under conditions of enhanced UV-B. Earlier research showed that willow chemical responses to enhanced UV-B were more clone-specific than species-specific,[79] again illustrating the importance of genetic variability in predicting responses to climate change. Studies of interactions between plant responses and insect herbivores under increased UV-B for two phytochemically different willow species (*Salix myrsinifolia* and *S. phylicifolia*) showed no direct effects on plant "quality" but increased insect abundance on both willow species, as well as increased numbers of a specialist herbivore on *S.* myrsinites.[21] Changes in insect abundance can have significant direct effects on habitat biodiversity, but they can also have significant effects on biodiversity at other trophic levels, through effects on insect-eating predators such as carnivorous insects and birds.[80] Plants browsed by mammalian herbivores can also suffer increased insect herbivory,[81] again demonstrating the importance of interactions more than direct single-factor effects in driving habitat responses to perturbation. Insects are likely to respond more quickly to climatic change than are the host plants. This could have severe impacts on the plants if insect growth and abundance increases under elevated temperatures and host-plant growth does not respond as quickly.[82] For example, increased winter temperatures have been found to lead to an upward altitudinal shift in the range of the pine processionary moth *Thaumetopoea pityocampa* in the Spanish Sierra Nevada, such that its range now overlaps with relic populations of *Pinus sylvestris* ssp. *nevadensis*, which have subsequently suffered from severe defoliation.[83] In view of their fast response rates, insect herbivores have actually been suggested as highly sensitive sensors of changing temperatures in arctic and alpine environments.[84] They could thus be used as early warning indicators of increased risk to key plant species of biodiversity importance (for example, *Dryas octopetala* and *Salix lanata*).

3.3   Fragmentation and Isolation

One of the most obvious characteristics of marginal habitats, particularly those which are strongly climatically limited, is that they tend to have a roughly linear distribution within a landscape. As described, both sub-arctic willow and mountain pine scrub inhabit the zone between the climatic limits of forest growth and the open sub-alpine meadow/dwarf shrub communities. Although in arctic systems this zone can be wide, in many upland areas it means that at any one site they inhabit a relatively narrow altitudinal band. This linearity of distribution, as well as the limitations of soil type, *etc.*, produces a spatial pattern that is highly vulnerable to fragmentation of the habitat. This spatial pattern, combined with their vulnerability to several key environmental change drivers, gives marginal habitats a high conservation profile.

The causes of fragmentation may be varied. We have discussed in Section 2 how land-use change, in particular grazing, can lead to habitat fragmentation by restricting the scrub species to inaccessible (and spatially fragmented) areas such as cliffs. Assuming that climate change more often will lead to an upslope movement of our two marginal scrub habitats, increasingly rugged topography at higher altitudes might also increase habitat fragmentation. For example, a fine-scale model of alpine species range-shifts in response to climate change predicted the restriction of some species to certain micro-habitats, such as hollows and the sheltered sides of ridges.[85] In the case of montane scrub vegetation, if increasing temperature forces an upward shift of the habitat, high wind speeds (and lack of snow-lie in such exposed areas) may also restrict the scrub species to sheltered hollows. Combining this with the factors discussed above which regulate the response of both the upper and lower limits of these scrub habitats to climate change, it is probable that whilst some patches are able to track a changing climate or persist *in situ*, others (due to local variation in topography and the composition and function of neighbouring communities) will not. Figure 5 demonstrates in a simplified manner how both grazing and climate change might lead to the same three possible outcomes (fragmentation, survival intact or extinction) for marginal upland scrub habitats.

Exploitation of the landscape for purposes other than agriculture may also lead to fragmentation of these systems. For example, it has been proposed that the existing fragmented populations of mountain pine in the Bulgarian mountains might be remnants of a previously large range that existed in the late Tertiary, and that in particular areas fragmentation of mountain pine distributions has resulted from harvesting of the trees for charcoal and firewood, as

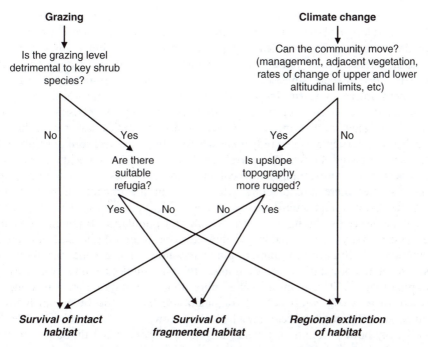

Figure 5 Possible outcomes of grazing and climate change (fragmentation, survival intact or extinction) for marginal upland scrub habitats.

well as alpine pasturing.[69] Charcoal and oil production, and copper and iron-ore mining, as well as the impact of cattle and sheep grazing (both directly through herbivory and also through clearance of thicket to create alpine pastures), have also significantly impacted mountain pine thickets in many areas since the 16th–17th centuries.[34] Currently, however, mineral extraction is not considered to be a major issue threatening this habitat. Tourism in mountain areas is also a cause of concern. For example, significant negative impacts of trampling on mountain pine habitats in the Vysoké Tatry Mountains have been found.[32] Increased tourism is not the only concern, but also potential changes in the type of tourism. In Swedish mountains, for example, there has been a notable shift away from self-reliant travel to more organised mechanised tourist activities, *e.g.* skidoo use and downhill skiing rather than hiking, cross-country skiing and camping. The former activities tend to create more localised, often damaging pressure than the latter.[86] However, whilst tourism can locally damage these marginal habitats, their rich landscape value (see Figures 1–3) can also act as attractants which bring tourists, and often much-valued finance, into an area. This can also play a fundamentally important role where a lack of awareness of the condition of a particular threatened habitat is an issue hampering its protection; publicising a habitat and its beauty, *etc.*, can directly lead to public action to lobby for greater protection.[9] The importance of this approach cannot be overlooked.

3.3.1 Isolation and Genetics. One of the concerns associated with fragmentation and long isolation of species is the impact on their genetics. The common prediction is that isolation leads to reduced genetic variability and therefore reduced capacity to respond to future changes/perturbations. For example, it was found that increasing isolation of *Fraxinus excelsior* populations at their northern range limit (in fragmented populations on Finnish islands) was associated with decreasing genetic variability, interruption of gene flow and increasing genetic differentiation.[2] However, not all studies have found this to be the case, particularly in windy areas where both pollen and seed of scrub and tree species have been found to travel very long distances.[87] Considerable effort has been put into examining the ecological consequences of rarity, and from these studies there is conflicting evidence about the impacts of population fragmentation on genetic diversity at the species level.[19,88,89] The equivocal results concerning the impacts of isolation and population fragmentation on genetic diversity are reflected in genetic studies of mountain pine and sub-arctic willow scrub. Studies of mountain pine have shown that genetic variability of some isolated populations is indeed lower than in samples from more extensive populations.[69,90] The greatest level of inbreeding in mountain pine was found[69] where there was greatest habitat fragmentation (*i.e.* in the Bulgarian mountains), which might be accentuated by its low growth form and a high degree of near-neighbour pollination. Single-site studies also indicated the same genetic effects. For example, a study of a population of 30 individuals of mountain pine in a peat bog site found only two genetically unique individuals – effectively two clonal populations[91] – which might result from a high degree of within-stand clonal layering, as discussed above. A relatively low gene flow between small peat bog populations has been attributed to their geographic isolation.[90] The Scottish Montane Willow Research Group[19] developed molecular tools to differentiate willow genotypes, with the aim of quantifying genetic diversity in the remaining Scottish populations of sub-arctic willow scrub and assessing the proportion of sexual *versus* asexual reproduction taking place. In contrast to the above research on mountain pine, the authors found no evidence for the anticipated loss of genetic diversity due to fragmentation and isolation of sub-arctic willows. They found that *S. lanata* and *S. lapponum*, for example, showed > 90% unique genetic individuals. They also found no differences in genetic diversity of rare versus more common willow species, poor correlations between population size and genetic diversity and relatively low population genetic differentiation, all indicating either high levels of gene flow (and ongoing sexual reproduction) or limited divergence "since fragmentation" (the authors have assumed that this habitat was once much more widespread in the UK).

A study across several different countries within mountain pine range[92] found no notable reductions in the degree of trait variation in isolated populations, indicating that although genetic variability can be reduced, morphological variability is not necessarily reduced by fragmentation and isolation, particularly if the isolated populations are locally abundant. One of the major research challenges, therefore, is to understand the implications of fragmentation for the genetics of populations of different species and to determine whether or not any such differences really have negative consequences for the populations involved. The potential consequences of isolation must also be considered within the context of the pattern of isolation and the rate of turnover of populations. If populations are small but stable, with low death rates, then the existing genetic diversity is likely to be maintained for a considerable period (as in the fragmented sub-arctic willow populations). The link between genetic diversity and individual plasticity (and the capacity to cope with a fluctuating environment) also needs to be carefully considered. Alpine and montane species are well known for their considerable individual longevity. For example, single clones of *Carex curvula* have survived for possibly 4000 years[93] and some *Pinus longaeva* trees in the White Mountains, California, have also been calculated to be over 4500 years old.[94]

3.3.2 Isolation and "Allee" Effects. In addition to potential genetic consequences of fragmentation and isolation, "allee effects" – defined as positive relationships between fitness and increasing population size in small populations[95] – can also operate through a number of mechanisms. Reproductive mechanisms, for example, may become less effective at low densities, especially those dependent on insect (as opposed to wind) pollination and outcrossing. Reduced sexual reproductive success was found in an invading front of *Spartina anglica* in an American salt marsh ecosystem until clonal growth increased

the local density of individuals.[96] The negative impact of isolation on reproductive success is a particular problem for the montane willows, which are
dioecious.[97,98] Added to this, some species are wholly dependent on insects for
pollination (*e.g. Salix lanata*) which further reduces their likelihood of pollination in isolated marginal habitats. A comparison of *S. lanata*, with its single
pollination strategy, and *S. lapponum*, a species with 50:50 insect:wind pollination, showed clear advantages for the latter species in alpine areas where
insect pollinator activity tends to be infrequent as wind speeds are generally
high.[99] Insect dependence on specific species is well known, but dependence on
female catkins has also been shown for two species of jumping lice on *S.
lapponum*,[82] further exacerbating the effects of isolation on biodiversity.

4   Managing Biodiversity in Marginal Habitats

So far we have described marginal habitats in general and in more detail for our
case-study mountain pine and sub-arctic willow habitats. We have explored, in
particular, the possible impacts of two key environmental change drivers,
climate change and land-use change, and one of the main knock-on effects of
these drivers for our case study habitats, *i.e.* habitat fragmentation. Although
there is uncertainty in predicting exactly where such impacts are likely to be of
particular importance, it is clear that some areas will need active management
in order to protect these habitats and in some instances this will demand further
research activity on which to base management decisions and perhaps even
policy change (as illustrated in Figure 4). In this section we first discuss the
policy framework that currently regulates biodiversity conservation and how
this relates to our two case studies, then discuss some of the research needs and
management actions required to conserve the biodiversity of these marginal
upland scrub habitats.

4.1   Policy Context

As discussed elsewhere in this book, the management of biodiversity in EU
countries is strongly regulated by European biodiversity policy, for example the
1979 EC Birds Directive (Council Directive 79/409/EEC) and the 1992 EC
Habitats Directive (Council Directive 92/43/EEC). Signatory nations to the
Habitats and Birds Directives have an obligation to protect particular listed
species and habitats and maintain the environment in a "favourable" state.
Research and management activities aimed at promoting the conservation of
biodiversity need to take account of, and work within, this policy framework,
as the framework sets priorities for conservation organisations.

   The Natura 2000 network, a Europe-wide network of nature conservation
sites aimed at protecting priority habitats and species, was initiated through the
Habitats Directive. Other key European policy developments include the
Convention on Biological Diversity and the commitment at the Gothenburg
Council by the EU Heads of State and Governments to "halt the decline of

biodiversity by 2010". However, despite such commitments, recent reviews have highlighted continued declines in Europe's biodiversity and they call for further action to conserve biodiversity.[100] In response, the Commission has recently produced a communication called "Halting the loss of biodiversity by 2010 – and beyond", which lays out a strategy for dealing with these problems and concerns.[40]

So how does this high-level policy relate to the habitats that we have been considering in this chapter? As mentioned above, the European Commission's Habitats Directive lists habitats and species (Annex 1 and 2, respectively) within the EU which are considered rare, threatened or deteriorating. There is a statutory requirement for member states to take appropriate measures (usually through creation of a network of SACs, *i.e.* Special Areas for Conservation) to maintain these habitats/species at, or restore them to, "favourable conservation status", which is achieved when:

- the natural range and areas covered are stable or increasing
- the specific structure and functions necessary for long-term maintenance exist and are likely to continue to do so for the foreseeable future
- the conservation status of the "typical" species is favourable (*i.e.* the population is being maintained as viable, the natural range is not reducing and there is sufficient habitat to maintain the population long term).

Sub-arctic willow scrub is an Annex 1 habitat, but there is no European-wide protection for mountain pine scrub in general, probably because the habitat is widespread and relatively abundant within its core areas. However, sub-categories of mountain scrub that include mountain pine are listed as priority habitats, in particular the Annex 1 habitat mountain pine scrub with *Rhodo-dendron hirsutum* (*Mugo-Rhododendretum hirsuti*) or other *Rhododendron* species, which is found throughout the mountain pine range (and is known as Alpenrosengebüsch). Furthermore, the Habitats Directive treats *P. mugo* subspp. *uncinata* as a full species and Annex 2 contains the habitat sub-alpine and montane *Pinus uncinata* forest (a priority habitat if on gypsum or lime-stone), which is found in the Alps, the Jura Mountains and the Pyrenees.

Alpine/montane scrub habitats therefore receive some level of *commitment* to protection under EU biodiversity conservation legislation. However, the effec-tiveness of this legislation has been questioned. The EC's recent Biodiversity Communication noted that, for a number of reasons, "While important progress has been made ... the pace and extent of implementation [of policies to halt biodiversity loss by 2010] has been insufficient". Why might this be so? First, implementation of European legislation occurs at a regional (national) level and the degree of implementation varies markedly between countries. For example, all the "main" areas of sub-arctic willow scrub are protected in the UK by the European Special Area for Conservation (SAC) status (*i.e.* EU status has been given to the largest known populations and site selection has taken account of floristic variations and the geographical range[101]). For its range in Sweden and Finland this habitat is generally protected through its

Annex 1 designation. In Norway this is not the case, although Norway has signed the UN Convention on Biological Diversity and the EU Convention on the Conservation of European Wildlife and Natural Habitats.[102] Emphasis is placed upon site-based designation processes such as the SACs and Natura 2000, but even following Natura Designation the Communication admits that some sites suffer from degradation, for example through land use and development, and that increasing population density is likely to have particular consequences in terms of increased exploitation in mountainous areas.

One of the major problems that may limit uptake of, and commitment to, biodiversity legislation is likely to be the failure of biodiversity in the policy arena. A recent review by the European Environment Agency[103] indicated that there was little evidence of national sustainability strategies, a key mechanism for environmental policy integration, being implemented. Failure of biodiversity legislation in the policy arena may result from the perceived cost-benefit ratio of biodiversity conservation measures. Although considerable effort has gone into attempting to cost the more ephemeral aspects of biodiversity (for example aesthetic value), it has been argued[104] that calculating the more tangible economic value of biodiversity, such as the value of ecosystem services provided by habitats (which may support in some cases up to 40% of a country's economic production[105]), would strengthen the case for biodiversity conservation (although a counter-argument also exists[106]). Some studies have attributed a potential direct economic value to these marginal scrub systems through their roles in slope stabilisation and avalanche protection[34] and as components of systems that act as natural filters for drinking water production (S. Dullinger – *personal communication*).

These are generalities which might apply to many habitats. However, there are also some specific issues that relate to marginal scrub environments and the particular threats that they face (as outlined above). First, there is the issue of the static nature of the designation process. Much of the policy behind these site-orientated approaches was initiated at a time when the significant likely impacts of climate change on biodiversity were not a major component of conservation philosophy. The site-orientated nature of current conservation approaches has been criticised in a major recent review of the likely impacts of climate change in Arctic environments[107] because it does not allow for species turnover within sites or promote the movement of species through the landscape. In the case of the marginal upland scrub habitats, we have discussed how the ability of these species to track an appropriate "climate window" will be vital in determining their survival during climate change. The creation of protected areas with sufficient environmental amplitude to cover not only existing ranges, but also potential future ranges or areas that are sufficiently well connected, will be vital for their survival during climate change. The recent Biodiversity Communication recognises the need for better integration between protected areas, for example "securing coherence of the Natura 2000 network", but questions remain concerning the impact of habitat fragmentation (and hence the degree of connectivity within the landscape) on the capacity of particular species and communities to track climate.[108]

Second, we have already shown that there may be some fundamental constraints on the movement of these species and communities in response to climate change. As a consequence it may be necessary to take active steps to ensure that species occur at sites which are likely to represent favourable future climatic conditions, *i.e.* through translocation. However, it has been suggested that the protection afforded to some species through the legislation may place limits on the possibility of management techniques such as transplantation being applied. The EU SAC rulings relating to favourable conservation status (as interpreted in the UK) make it difficult to allow a net loss of a habitat on protected sites, for example, so the expansion of one designated habitat can be difficult if it is at the expense of other Annex 1 habitats. This can often be the case in montane areas where several key habitats are protected under EU legislation. In addition, because some marginal habitats are often classed as "transitional" in nature, they can suffer a lack of protection by falling between specific conservation management incentives. For example, in the UK, grants for expansion of native forest have only recently included tree-line woodland under tightly controlled conditions, and montane scrub has only very recently been included in a broad Montane Habitat Action Plan.

It is not only conservation policy which impacts upon these habitats. We have discussed the importance of grazing for these systems and in many areas, for example alpine grassland systems, we have seen how changes in land management practice can have significant consequences for habitat distributions. Therefore, changes in policy other than conservation policy will be directly relevant to future changes in the conservation and biodiversity status of these habitats. Current reform of European farming policy, for example, is likely to have significant impacts, as will policies that affect other exploitation processes in these systems, *e.g.* tourism. In the light of previous criticism concerning the lack of cross-sectoral uptake of actions for biodiversity conservation,[103] the Biodiversity Communication recognises the need for a more integrated approach, including "optimising the use of available measures under the reformed CAP, notably to prevent intensification of abandonment of high-nature-value farmland, woodland and forest" and "ensuring that community funds for regional development benefit, and do not damage, biodiversity".

It is clear, therefore, that future conservation of these habitats is not simply a case of developing appropriate management regimes supported by targeted research, but must also involve development, application and regular reassessment of appropriate policy approaches in a number of sectors.

4.2 Research Priorities

It is almost a truism that a document written by researchers will point out the need for further research. However, management and policy development do need good research results as a basis for decision making and we have seen above that there are many issues relating to marginal habitats and the consequences of climate or other change which are still poorly understood. We propose four main areas below that require targeted research activities which we consider to be essential to the management of the type of marginal habitats that we have focused upon in this chapter (though, of course, this list is not exhaustive). Because of the nature of marginal habitats, these are also applicable to a wider range of systems than our case-study willow and mountain pine communities.

- *Positive and negative impacts of fragmentation on biodiversity.* The montane scrub communities present excellent test cases for such studies. What is needed is a clearer understanding of how detrimental fragmentation can be, both to the scrub species themselves and to those species that are dependent upon them. This would enable understanding of: (1) the amount of time that is available before isolation becomes critical (genetically, physiologically or otherwise) to a specific area/habitat and (2) the optimum way to invest limited biodiversity conservation resources in different areas. A good example of the need for clear research outputs and better science/policy-maker communication, for example, is the re-examination of the commonly accepted belief (discussed above) that fragmentation is always "a bad thing" for genetic diversity.
- *Scenario modelling of predicted net effects of key drivers on expansion and contraction of upper and lower range margins under different conditions.* This is crucial to predict the resilience of a habitat to change, *i.e.* the likelihood of habitat expansion, compression or fragmentation under predicted future conditions in different areas. As discussed, biodiversity drivers have strong, interactive effects and so such modelling should incorporate land-use as well as climate-change scenarios. The FORET/ JABOWA-type forest succession model example given in Section 3.2 above[71] is a good example of this type of approach applied to one of our case study marginal habitats. Clearly, underpinning experimental research results in all the key areas outlined earlier are crucial for model development and parameterisation.
- *Estimating the true economic value of different marginal habitats.* Economic values have often been overlooked in the past when their value is not direct, *e.g.* timber or other similar output potential, but this area of valuation is becoming increasingly recognised as important, although biodiversity valuation is still in its relative infancy and the use of some established valuation techniques have been the subject of critical debate.[109–112] To take one of our case study habitats, the role of mountain pine scrub in avalanche protection and slope stabilisation is clear, but other "economic" benefits from protecting these systems have been little considered. As pointed out by the EC Biodiversity Communication "An important driver of biodiversity loss, along with population growth and growing *per capita* consumption, is the failure of conventional economics to realise the economic value of natural capital and ecosystem services", but should we perhaps also promote, as suggested by McCauley,[106] the ethical basis for conservation of these systems, and would such an

argument actually provide effective support for their conservation? This type of research can only be undertaken in an interdisciplinary manner, combining skills from the ecological, economic and social sciences.

- *Prioritising areas for conservation action.* Should habitats of conservation concern be protected across their whole range or should protection (usually involving limited resources) focus on specific areas? If the latter, as is usually the case, how best should we select target areas for biodiversity conservation? Should we protect only core areas, where we might predict greatest likelihood of success, or should we focus on fragmented areas which are currently most vulnerable to change and which might provide "refugia" for species which are extremely rare in those areas? To date, the latter approach has been most common, but this is not necessarily the best use of resources. As well as the obvious scientific needs underlying these questions, such decisions require international integration of conservation effort to a much greater degree than exists at present. As with the above proposal, this research topic would need involvement of a range of scientific skills, involving as it does the target of achieving a difficult balance between science-driven and policy-driven priorities.

In addition, it is essential that the results of these research activities are properly communicated to policy makers. This is a notably weak point in research-result dissemination in general, but it does appear to be receiving significant current support, for example through mechanisms such as the European Platform for Biodiversity Research Strategy (EPBRS), which aims to be "a forum for scientists and policy makers to ensure that research contributes to halting the loss of biodiversity by 2010". Associated with increased attempts by scientists to disseminate the results of their research, it is perhaps also necessary for policy-makers to understand the limitations of research and the necessary caveats placed around research results by scientists. The bottom line is perhaps that although scientists can rarely provide definitive answers, this is still often the best information that policy makers will receive.

4.3   Management Action

Protection of biodiversity under threat is often split into immediate actions required and longer-term management needs, and the two may differ considerably. If we take sub-arctic willow scrub as an example, immediate actions may include propagation by cuttings to ensure survival of genetic material, where fragmentation and isolation has resulted in little or no seed production and immediate danger of local extinction.[19,97,98] Medium- to longer-term action may involve propagation and transplanting (of both sexes) to reduce isolation and facilitate sexual reproduction.[19,97,98] However, the latter actions will only succeed if the causes of the fragmentation have been addressed. For example, if heavy grazing is the cause then transplanting without reducing or protecting from grazing will fail. Similarly, if climate change has rendered conditions around current "refugia" unsuitable for growth of this habitat (*e.g.* reductions in winter snow cover), then attempts at expansion in the same areas will also fail.

The above types of approach are generally classed as "high intervention". As well as the limitations to this approach from designations such as Natura 2000, as discussed above, in some areas there is a presumption against this in favour of allowing "natural processes" to determine the future status of the habitat[113] and therefore the above options are not always possible in certain areas. This is clearly a difficult issue where a habitat is severely threatened; do we take a "low intervention" approach and take the risk of local extinction, or do we actively intervene to ensure its local survival? This is probably one of the most controversial questions in biodiversity conservation management. Mardon[98] gives a good, considered debate on the above issues for montane willow scrub.

In areas where the marginal habitat requiring protection is not as fragmented or immediately threatened, more general, less interventionist, protective management is often the only action taken. This may take the form of management agreements with land owners to reduce grazing or manage tourism, for example, and/or creation of National Park areas, which may do the same thing over a bigger area. For example, the establishment of National Parks in (as was) Czechoslovakia limited human impacts (livestock grazing) and enabled mountain pine regrowth to occur.[34] Moreover, the management strategy of a new National Park in Scotland included, for the first time, explicit protection of montane scrub.[114] However, as above, National Park-type protection will only work if the conditions created within the park are suitable for protection or expansion of the targeted habitat. Grazing has already been mentioned, but in the case of predicted climate change scenarios the area will also need to include potential "expansion" areas under predicted future as well as current climatic conditions. This may not always be feasible, especially for high-altitude communities which may have nowhere locally to expand into, should conditions become significantly warmer. In addition, competitive exclusion from surrounding habitats may prevent expansion, as discussed earlier, and this would require careful initial planning and subsequent monitoring to ensure success.

Protection and expansion targets set by conservation policies, as discussed earlier, generally drive management action, but for all the above management options it is essential that strategic planning is carried out to ensure the best use of what are normally limited resources for biodiversity conservation. Conservation priorities must consider both current and future potential and, as discussed above, in many cases this requires differential management to ensure short- and longer-term survival.

Acknowledgements

We are grateful to the Scottish Executive Environment and Rural Affairs department for funding AJH and RB. Thanks to Diana Gilbert for valuable discussions and suggestions on earlier drafts. David Mardon kindly provided the photograph for Figure 2. Thanks to Robert Kanka for providing information about *P. mugo* in Slovakia, to Stefan Dullinger for providing valuable comments on the chapter and to Lucio Di Cosmo for translating his manuscript into English. We are also grateful to ALTER-Net colleagues for useful discussions about drivers of biodiversity change.

References

1. S. A. O. Cousins, *Biol. Cons.*, 2006, **127**, 500–509.
2. A. M. Höltken, J. Tähtinen and A. Pappinen, *Silvae Genetica*, 2003, **52**, 206–212.
3. E. W. Seabloom, E. T. Borer, V. L. Boucher, R. S. Burton, K. L. Cottingham, L. Coldwasser, W. K. Gram, B. E. Kendall and F. Micheli, *Ecol. Applic.*, 2003, **13**, 575–592.
4. P. Choler, B. Erschbamer, A. Tribsch, L. Gielly and P. Taberlet, *Proc. Natl. Acad. Sci. USA*, 2004, **101**, 171–176.
5. S. Lavergne, W. Thuiller, J. Molina and M. Debussche, *J. Biogeog.*, 2005, **32**, 799–811.
6. T. Herben, Z. Munzbergova, M. Milden, J. Ehrlen, S. A. O. Cousins and O. Eriksson, *J. Ecol.*, 2006, **94**, 131–143.
7. L. Poldini, G. Oriolo and C. Francescato, *Plant Biosystems*, 2004, **138**, 53–85.
8. J. J. Camarero, E. Gutiérrez and M. -J. Fortin, *Global Ecol. Biogeog.*, 2006, **15**, 182–191.
9. Scottish Natural Heritage, "Montane scrub. Natural heritage management", Scottish Natural Heritage, Battleby, 2000.
10. D. Gilbert, D. Horsfield and D. B. A. Thompson (eds), "The ecology and restoration of montane and subalpine scrub habitats in Scotland", *Scot. Nat. Herit. Rev.*, 1997, **83**, 1–128.
11. S. Dullinger, T. Dirnböck, R. Kock, E. Hochbichler, T. Englisch, N. Sauberer and G. Grabherr, *J. Ecol.*, 2005, **93**, 948–957.
12. D. N. McVean and D. A. Ratcliffe, "Plant Communities of the Scottish Highlands", HMSO, Edinburgh, 1962.
13. J. S. Rodwell (ed.), "British Plant Communities, Vol 1. Woodlands and Scrub", Cambridge University Press, Cambridge, 1991.
14. D. L. Jackson and C. R. McLeod (eds), "Handbook on the UK status of EC Habitats Directive interest features: provisional data on the UK distribution and extent of Annex I habitats and the UK distribution and population size of Annex II species, Revised 2002", *JNCC Report* 2000, **312**, 1–180.
15. E. Dahl, "Rondane: Mountain Vegetation in South Norway and its Relation to the Environment", Aschehoug, Oslo, 1956.
16. J. J. Hopkins and A. L. Buck, "The Habitats Directive Atlantic Biogeographical Region, Report of the Biogeographical Region Workshop,

Edinburgh, Scotland, 13–14 October 1994", *JNCC Report* 1995, **247**, 1–39.

17. K. P. Bland, P. F. Entwistle and D. Horsfield in "The ecology and restoration of montane and subalpine scrub habitats in Scotland", D. Gilbert, D. Horsfield and D. B. A. Thompson (eds), *Scot. Nat. Herit. Rev.*, 1997, **83**, 35–41.

18. S. R. Mortimer, A. J. Turner, V. K. Brown, R. J. Fuller, J. E. G. Good, S. A. Bell, P. A. Stevens, D. Norris, N. Bayfield and L. K. Ward, "The nature conservation value of scrub in Britain", *JNCC Report* 2000, **308**, JNCC, Peterborough.

19. Scottish Montane Willow Research Group, "Biodiversity, Taxonomy, Genetics and Ecology of Sub-arctic Willow Scrub", Royal Botanic Gardens, Edinburgh, 2005.

20. H. Roininen, K. Danell, A. Zinovjev, V. Vikberg and R. Virtanen, *Polar Biol.*, 2002, **25**, 605–611.

21. T. O. Veteli, R. Tegelberg, J. Pusenius, M. Sipura, R. Julkunen-Tiitto, P. J. Aphalo and J. Tahvanainen, *Oecologia*, 2003, **137**, 312–320.

22. L. Nagy in "Alpine Biodiversity in Europe", L. Nagy, G. Grabherr, C. Körner and D. B. A. Thompson (eds), Springer-Verlag, Berlin, *Ecological Studies*, 2003, **167**, 39–46.

23. A. Shevtsova, E. Haukioja and A. Ojala, *Oikos*, 1997, **78**, 440–458.

24. J. Olofsson, *Arct. Antarct. Alp. Res.*, 2004, **36**, 464–467.

25. O. Totland and J. Esaete, *Plant Ecol.*, 2002, **161**, 157–166.

26. R. Marriott in "The ecology and restoration of montane and subalpine scrub habitats in Scotland", D. Gilbert, D. Horsfield and D. B. A. Thompson (eds), *Scot. Nat. Herit. Rev.*, 1997, **83**, 41–44.

27. N. A. MacKenzie, "Low Alpine, Sub-alpine and Coastal Scrub Communities in Scotland", Highland Birchwoods, Munlochy, 2000.

28. K. I. Christensen, *Nordic J. Botany*, 1987, **7**, 383–408.

29. G. Lehner and C. Lutz, *J. Plant Physiol.*, 2003, **160**, 153–166.

30. A. Berkutenko, *Int. Dendrol. Soc. Yearbook 1992*, 1993, 41–46.

31. J. Nascimbene, G. Caniglia and M. D. Vedove, *Cryptog. Mycolog.*, 2006, **27**, 185–193.

32. F. Kubíček, L. Šomšák, V. Šimonovič, E. Majzlanová, I. Háberová and V. Rybárska, *Folia Geobot. Phtytotx.*, 1983, **18**, 363–387.

33. T. Dirnböck and G. Grabherr, *Mountain Res. Dev.*, 2000, **20**, 172–179.

34. M. Jodłowski in: "Global Change in Mountain Regions", M. F. Price (ed.), Sapiens Publishing, Duncow, Scotland, 2006, 186–187.

35. I. Stefanova, N. Ognjanova-Rumenova, W. Hofmann and B. Ammann, *J. Paleolimnol.*, 2003, **30**, 95–1.

36. S. Tonkov and E. Marinova, *Holocene*, 2005, **15**, 663–671.

37. A. A. Ali, M. Martinez, N. Fauvart, P. Roiron, G. Fioraso, J. -L. Guendon, J. -F. Terral and C. Carcaillet, *Plant Biol. Pathol.*, 2006, **329**, 494–501.

38. S. Dullinger, T. Dirnböck and G. Grabherr, *J. Ecol.*, 2004, **92**, 241–252.

39. European Environment Agency, http://glossary.eea.europa.eu/EEAGlossary/D/DPSIR, 14th Sept. 2006.
40. European Commission, "Communication from the Commission. Halting the Loss of Biodiversity by 2010 – and Beyond: Sustaining Ecosystem Services for Human Well-being", EC, Brussels, 2006.
41. L. Nagy, G. Grabherr, C. Körner and D. B. A. Thompson in "Alpine Biodiversity in Europe", L. Nagy, G. Grabherr, C. Körner and D. B. A. Thompson (eds), Springer-Verlag, Berlin, *Ecological Studies*, 2003, **167**, 453–464.
42. B. Sveinbjörnsson, *Ambio*, 2000, **29**, 388–395.
43. J. Grace, F. Berninger and L. Nagy, *Ann. Bot. London*, 2002, **90**, 537–544.
44. P. Balaź and J. Mindźś, *Ekológia (Bratislava)*, 2004, **23**, 1–12.
45. L. Kullman, *J. Ecol.*, 2002, **90**, 68–77.
46. S. Hale, C. P. Quine and J. C. Suárez, *Scot. Forest*, 1998, **52**, 70–76.
47. R. Geiger, "The Climate Near the Ground", Harvard University Press, Cambridge, MA, USA, 1965.
48. C. Körner, "Alpine Plant Life", Springer, New York, 1999.
49. J. Grace, *Bot. J. Scotl.*, 1997, **49**, 223–236.
50. C. Körner, *Oecologia*, 1998, **115**, 445–459.
51. J. Senn, *Eur. J. Forest Pathol.*, 1999, **29**, 65–74.
52. L. Di Cosmo, "Considerazione sull'esistenza delle annate di pasciona nel *Pinus mugo* mediante l'analisi dendroecologica in una stazione della maiella (1)", *L'Italia Forestale e Montana n. 3*, University of Tuscia, Italy, 2003.
53. N. M. Schmidt, C. Baittinger and M. C. Forchhammer, *Arct. Antarct. Alp. Res.*, 2006, **38**, 257–262.
54. A. Tolvanen and G. H. R. Henry, *Can. J. Bot.*, 2001, **79**, 711–718.
55. S. Hättenschwiler and C. Körner, *J. Veg. Sci.*, 1995, **6**, 357–368.
56. M. Scheffer, S. Carpenter, J. A. Foley, K. Folke and B. Walker, *Nature*, 2001, **413**, 591–596.
57. J. P. Theurillat and A. Schlussel, *Phytocoenologia*, 2000, **31**, 439–456.
58. J. Silvola and U. Ahlholm, *Oikos*, 1993, **67**, 227–234.
59. M. H. Jones, C. Bay and U. Nordenhall, *Global Change Biol.*, 1997, **3**, 55–60.
60. T. O. Veteli, J. Kuokkanen, R. Julkunen-Tiitto, P. J. Aphalo, H. Roininen and J. Tahvanainen, *Glob. Change Biol.*, 2002, **8**, 1240–1252.
61. S. Turtola, M. Rousi, J. Pusenius, K. Yamaji, S. Heiska, V. Tirkkonen, B. Meier and R. Julkunen-Tiitto, *Glob. Change Biol.*, 2005, **11**, 1655–1663.
62. S. Turtola, M. Rousi, J. Pusenius, K. Yamaji, V. Tirkkonen, B. Meier and R. Julkunen-Tiitto, *Environ. Exp. Bot.*, 2006, **56**, 80–86.
63. J. Moen in "Global Change in Mountain Regions", M. F. Price (ed.), Sapiens Publishing, Duncow, Scotland, 2006, 187–188.
64. T. Dirnböck, S. Dullinger and G. Grabherr, *J. Biogeog.*, 2003, **30**, 401–417.
65. K. Tape, M. Sturm and C. Racine, *Glob. Change Biol.*, 2006, **12**, 686–702.

66. S. Dullinger, T. Dirnböck and G. Grabherr, *Arct. Antarct. Alp. Res.*, 2003, **35**, 434–441.
67. E. A. Gage and D. J. Cooper, *Can. J. Bot.*, 2005, **83**, 678–687.
68. R. Gill in "Large Herbivore Ecology and Ecosystem Dynamics", K. Danell, R. Bergström, P. Duncan and J. Pastor (eds), Cambridge University Press, Cambridge, 2006, 170–202.
69. G. T. Slavov and P. Zhelev, *Can. J. For. Res.*, 2004, **34**, 2611–2617.
70. A. Stanisci, G. Pelino and C. Blasi, *Biodivers. Conservat.*, 2005, **14**, 1301–1318.
71. F. Kienast, J. Fritschi, M. Bissegger and W. Abderhalden, *Forest Ecol. Manag.*, 1999, **120**, 35–46.
72. F. J. G. Mitchell and K. J. Kirby, *Forestry*, 1990, **63**, 333–353.
73. K. A. Holland, W. C. Leininger and M. J. Trlica, *Rangeland Ecol. Manage.*, 2005, **58**, 148–154.
74. A. J. Hester, M. Bergman, G. R. Iason and R. Moen in "Large Herbivore Ecology and Ecosystem Dynamics", K. Danell, R. Bergström, P. Duncan and J. Pastor (eds), Cambridge University Press, Cambridge, 2006, 97–141.
75. S. Matouch, B. Arroyo, J. Forster, P. Scott Jones, A. Kuzniar, R. Mitchell, M. Rebane, S. Redpath, P. Rose, C. Scheidegger, D. Soier and S. Stoll-Kleeman in "Conflicts between human activities and the conservation of biodiversity in agricultural landscapes, grasslands, forests, wetlands and uplands in Europe. A report of the BIOFORUM project August 2003", J. Young, P. Nowicki, D. Alarad, K. Henle, R. Johnson, S. Matouch, J. Niemelä and A. Watt (eds), Centre for Ecology & Hydrology, Banchory, Scotland, 2003, 123–144.
76. S. R. Halloy and A. F. Mark, *Arct. Antarct. Alp. Res.*, 2003, **35**, 248–254.
77. G. R. Miller, R. P. Cummins and A. J. Hester, *Scot. Forestry*, 1998, **52**, 14–19.
78. C. Stolter, J. P. Ball, R. Julkunen-Tiitto, R. Lieberei and J. U. Ganzhorn, *Can. J. Zool.*, 2005, **83**, 807–819.
79. R. Tegelberg, T. Veteli, P. J. Aphalo and N. Julkunen-Tiitto, *Basic Appl. Ecol.*, 2003, **4**, 219–228.
80. M. Sipura, *Oecologia*, 1999, **121**, 537–545.
81. J. Olofsson and J. Strengbom, *Oikos*, 2000, **91**, 493–498.
82. J. K. Hill and I. D. Hodkinson, *Ecol. Entomol.*, 1995, **20**, 237–244.
83. J. A. Hódar and R. Zamora, *Biodivers. Conservat.*, 2004, **13**, 493–500.
84. I. D. Hodkinson and J. Bird, *Arct. Alp. Res.*, 1998, **30**, 78–83.
85. M. Gottfried, H. Pauli, K. Reiter and G. Grabherr, *Diversity and Distributions*, 1999, **5**, 241–251.
86. P. Fredman and T. A. Heberlein in "Mountains of Northern Europe: Conservation, Management, People and Nature", D. B. A. Thompson, M. F. Price and C. A. Galbraith (eds), TSO Scotland, Edinburgh, 2005, 203–212.
87. R. A. Ennos, *Evolution*, 2005, **59**, 979–990.

88. J. D. Karron, *Evolutionary Ecol.*, 1987, **1**, 47–58.
89. M. A. Gitzendanner and P. S. Soltis, *Am. J. Bot.*, 2000, **87**, 783–792.
90. W. Prus-Glowacki, A. Baczkiewicz and D. Wysocka, *Acta Biol. Cracov. Bot.*, 2005, **47**, 53–59.
91. A. Bączkiewicz and P. Prus-Głowacki, *Acta Societas Botanicorum Poloniae*, 1997, **66**, 79–82.
92. K. Boratyńska, K. Marcysiak and A. Boratyński, *Bot. J. Linn. Soc.*, 2005, **147**, 309–316.
93. T. Steinger, C. Körner and B. Schmid, *Oecologia*, 1996, **105**, 94–99.
94. Anonymous, *Tree-Ring Bulletin*, 1958, **22**, 2–6.
95. W. C. Allee, "Animal Aggregations: a Study in General Sociology", University of Chicago Press, Chicago, 1931.
96. H. G. Davis, C. M. Taylor, J. C. Civille and D. R. Strong, *J. Ecol.*, 2004, **92**, 321–327.
97. D. Mardon in "The ecology and restoration of montane and subalpine scrub habitats in Scotland", D. Gilbert, D. Horsfield and D. B. A. Thompson (eds), *Scot. Nat. Herit. Rev.*, 1997, 83, 65–74.
98. D. Mardon in "Montane Scrub: the Challenge Above the Treeline", D. Gilbert (ed.), Highland Birchwoods, Munlochy, 2002, 22–31.
99. O. Totland and M. Sottocornola, *Am. J. Bot.*, 2001, **88**, 1011–1015.
100. Message from Malahide "Halting the Decline of Biodiversity – Priority Objectives and Targets for 2010", Stakeholders' Conference "Biodiversity and the EU – Sustaining Life, Sustaining Livelihoods" Grand Hotel, Malahide, Ireland, 2004.
101. JNCC website, SAC selection, http://www.jncc.gov.uk/protectedsites/ sacselection/ habitat.asp?FeatureIntCode=H4080, 10 Sept. 2006.
102. UN Convention on Biological Diversity, http://conventions.coe.int/ treaty/en/Treaties/Html/104.htm, 14 Sept. 2006.
103. European Environment Agency, "Environmental Policy Integration in Europe: state-of-play and an evaluation framework", Technical report No. 2/2005, EEA, Copenhagen, Denmark, 2005.
104. R. T. Watson, *Phil. Trans. Roy. Soc. B.*, 2005, **360**, 471–477.
105. Convention on Biological Diversity, "Integration of biodiversity considerations in the implementation of adaptation activities to climate change at the local, subnational, national, subregional and international levels", Note from the Executive Secretary to the AHTEG on Biodiversity and Adaptation to Climate Change, 4th July 2005, UNEP/CBD/AHTEG-BDACC/1/2.
106. D. J. McCauley, *Nature*, 2006, **443**, 27–28.
107. ACIA, "Arctic Climate Impact Assessment", Cambridge University Press, Cambridge, 2004.
108. J. M. J. Travis, *Proc. Roy. Soc. Lond. B. Bio.*, 2002, **270**, 467–473.
109. P. A. L. D. Nunes and J. C. J. M. van den Bergh, *Ecol. Econ.*, 2001, **39**, 203–222.
110. C. Spash, *J. Econ. Psychol.*, 2002, **23**, 665–687.
111. I. Bräuer, *Agr. Ecosyst. Environ.*, 2003, **98**, 483–491.

112. M. Christie, N. Hanley, J. Warren, K. Murphy, R. Wright and T. Hyde, *Ecol. Econom.*, 2006, **58**, 304–317.
113. T. Clifford in "The ecology and restoration of montane and subalpine scrub habitats in Scotland", D. Gilbert, D. Horsfield and D. B. A. Thompson (eds), *Scot. Nat. Herit. Rev.*, 1997, **83**, 95–102.
114. A. Raven in "Montane scrub: the challenge above the treeline", D. Gilbert (ed.), Highland Birchwoods, Munlochy, 2002, 15–17.

Trends in Biodiversity in Europe and the Impact of Land-use Change

A. D. WATT, R. H. W. BRADSHAW, J. YOUNG, D. ALARD,
T. BOLGER, D. CHAMBERLAIN, F. FERNÁNDEZ-GONZÁLEZ,
R. FULLER, P. GURREA, K. HENLE, R. JOHNSON, Z. KORSÓS,
P. LAVELLE, J. NIEMELÄ, P. NOWICKI, M. REBANE, C.
SCHEIDEGGER, J. P. SOUSA, C. VAN SWAAY AND A. VANBERGEN

1 Introduction

The loss of biodiversity in Europe and elsewhere has been highlighted for several decades.[1–5] The scale and potential consequences of this loss has led to action to combat it, notably the Convention on Biological Diversity (CBD). Realising that current policies and action taken to conserve biodiversity were inadequate, the European Union at its 2001 summit meeting in Göteborg, Sweden, set the ambitious target to "protect and restore habitats and natural systems and halt the loss of biodiversity by 2010". A similar target was set by the CBD in 2002 "to achieve by 2010 a significant reduction of the current rate of biodiversity loss" and endorsed by the World Summit on Sustainable Development (WSSD) in 2002.

In this chapter, we discuss the available data on the loss of biodiversity in Europe, the probable causes of this loss, future threats to biodiversity, the policy response to these threats and recent research on measuring trends in biodiversity, with particular emphasis on detecting the impact of one major threat to biodiversity, change in land use. Nigel Boatman *et al.* present additional related and complementary material in Chapter 1 of this volume, with particular emphasis on farmland in the UK.

2 Biodiversity in Europe: Current Status

There are an estimated approximately 250 mammal, 500 bird, 70 amphibian, 200 reptile, 220 freshwater fish, 200 000 invertebrate and 12 500 plant species in

Issues in Environmental Science and Technology, No. 25
Biodiversity Under Threat
Edited by RE Hester and RM Harrison
© The Royal Society of Chemistry, 2007

Europe.[6] The numbers of species of some groups, like plants, birds and butterflies, are well known[7] but the biodiversity of other groups, such as many invertebrate taxa, is poorly understood. Likewise, although the distribution of some groups is well documented, particularly at the national scale,[8] there are no readily available sources of information on other taxa. For some groups of organisms, there are good data on their diversity and distribution but they have not been coordinated. Recent European initiatives to coordinate data on biodiversity include the Species2000[9] and Fauna Europea[10] projects. Globally, the Global Biodiversity Information Facility (GBIF)[11] coordinates biodiversity information, working in collaboration with existing programmes and with natural history museums and other organisations. The European Network for Biodiversity Information (ENBI)[12] and the European Invertebrate Survey[13] are among several major initiatives to coordinate information on Europe's biodiversity. Notable national initiatives, in some cases covering several taxa, include the Swedish Species Information Centre,[14] the Luomus project[15] in Finland, the Swiss Biodiversity Forum[16] and the biodiversity monitoring network in Hungary.[17] In the UK, biodiversity data are coordinated by the Biological Records Centre (BRC).[18] The BRC was established in 1964 as the national centre for the recording of freshwater and terrestrial biota, except birds for which the British Trust for Ornithology has major data holdings. The function of the BRC is to capture, manage, interpret and disseminate data on the past and present distributions of species at the geographical scale in the UK. The BRC archives currently contain >14 million records of $>12\,000$ UK species of plant, mammal, reptile, amphibian, fish, and 39 invertebrate groups ranging from spiders, beetles and butterflies to water-fleas, molluscs and annelids. These data have been used to map species' UK distributions at the $10\ \mathrm{km}^2$ scale.[19,20] In 2004, the National Biodiversity Network (NBN) Gateway[21] became fully operational. Founded by a consortium including the BRC, the NBN is the UK node of GBIF and brings together information on biodiversity from statutory agencies, national societies, local records centres and non-departmental government bodies. Over 20 million species records are currently accessible through the NBN from over 130 different datasets.

Although our knowledge of Europe's biodiversity is steadily increasing, it suffers from several problems in addition to the lack of coordination. One of the main problems is the patchy nature of the available data, not only because data from some countries or regions are scarce but also because of actual or potential recording biases within countries, a problem common throughout the world.[22] Data on biodiversity are frequently collected in areas where biodiversity is already known or thought to be high, leaving the biodiversity of some areas, particularly remote areas, poorly documented. Managed landscapes, particularly agricultural areas and managed forests, and urban areas are also relatively neglected. This problem results from the fact that the collection of data on biodiversity has rarely been planned so as to provide adequate biogeographical coverage. Much of the data have been collected by amateurs who are often not funded to collect data rigorously, although a high standard of systematic monitoring and survey of birds is now being achieved through volunteers in several countries. An even more serious problem is that few datasets provide useful information on temporal trends in biodiversity. There are notable exceptions, some of which are discussed below, but the lack of spatially and temporally comprehensive data on biodiversity is a serious impediment to quantifying biodiversity loss, understanding its causes and adequately responding to it.

3 Biodiversity in Europe: Information on Current Trends

There are many reports of biodiversity loss in Europe, most of which relate to the declining abundance of species or reduction in their distribution.[5,23,24] Since 1600, 16 species have been recorded as extinct in Europe, compared with 784 globally[25] (Table 1). However, IUCN list 142 as critically endangered, 143 as endangered, 425 as vulnerable, 27 as conservation-dependent and 223 as near-threatened.[25] National extinctions are well documented in Europe but several have led to successful reintroduction programmes.[26] Local loss of biodiversity has often been recorded too[27–30] but, considering the number of species and habitats in Europe, information on trends in biodiversity is extremely poor. Nevertheless, useful data on trends in biodiversity come from several sources.

3.1 Habitat Extent and Quality

Europe is extremely rich in habitats.[31] The CORINE classification of habitats for the EU Habitats Directive lists 58 different forest habitats alone.[32] Larsson *et al.*[32] present maps of 25 different forest types, many of which, including mixed oak forest and laurel forest, now cover a very small proportion of their potential extent. Data on the total area of forest in Europe show an expansion in forest cover in the last 30 years.[33] However, these statistics can mask decreases in areas of natural forest and increases in plantations of non-native species such as eucalyptus in Portugal.[34] Information on current trends in the extent of forest in Europe is provided by forest inventories and remote sensing.[35,36] Recent changes in the extent of many other habitats are also being recorded. The amount of semi-natural grassland in Europe has declined sharply in recent years;[6] between 1990 and 1998 it declined by 13% in the UK.[37] There has been a decrease in low-intensity farming systems, or "high nature value farmland", across Europe.[38,39] The loss of wetlands in Europe has been particularly dramatic, ranging from 60% in Denmark to 90% in Bulgaria since around the start of the twentieth century.[40]

Information on habitat quality is currently being assessed through various initiatives, such as the monitoring of habitats covered by Biodiversity Action Plans (BAPs) in the UK, with reporting every three years. However, more comprehensive monitoring of habitat quality will form part of the assessment of "favourable conservation status" of habitats in sites designated by the EU Birds and Habitats Directives. The favourable conservation status of a habitat

Table 1 Recent known extinctions of European species.[25]

Species	Order and family	Country and date of extinction
Astragalus nitidiflorus	Fabales Leguminosae	Spain
Belgrandiella intermedia	Mesogastropoda Hydrobiidae	Austria
Bythinella intermedia	Mesogastropoda Hydrobiidae	Austria
Chondrostoma scodrense	Cypriniformes Cyprinidae	Albania, Serbia and Montenegro
Telestes ukliva	Cypriniformes Cyprinidae	Croatia
Gallotia auaritae	Squamata Lacertidae	Spain
Graecoanatolica macedonica	Mesogastropoda Hydrobiidae	Greece *etc.* (1988–1992)
Haematopus meadewaldoi (Canary Islands Oystercatcher)	Charadriiformes Haematopodidae	Canary Is *etc.* (1980s)
Hydropsyche tobiasi (Tobias' Caddisfly)	Trichoptera Hydropsychidae	Germany
Leiostyla lamellosa (Madeiran Land Snail)	Stylommatophora Pupillidae	Madeira
Ohridohauffenia drimica	Mesogastropoda Hydrobiidae	Serbia and Montenegro
Pinguinus impennis (Great Auk)	Charadriiformes Alcidae	Iceland *etc.* (1850s)
Prolagus sardus (Sardinian Pika)	Lagomorpha Ochotonidae	France and Italy
Pseudocampylaea loweii	Stylommatophora Helicidae	Madeira
Radula visiniaca	Jungermanniales Radulaceae	Italy (1930s)
Siettitia balsetensis (Perrin's Cave Beetle)	Coleoptera Dytiscidae	France

is defined within the Habitats Directive as: "Its natural range and areas it covers within that range are stable or increasing, and the species structure and functions which are necessary for its long-term maintenance exist and are likely to continue to exist for the foreseeable future, and the conservation status of its typical species is favourable as defined [by when] population dynamics data on the species concerned indicate that it is maintaining itself on a long-term basis as a viable component of its natural habitats, and the natural range of the species is neither being reduced nor is likely to be reduced for the foreseeable future, and there is, and will probably continue to be, a sufficiently large habitat to maintain its population on a long-term basis". Unsurprisingly, the implementation of favourable conservation status monitoring is seen as a major challenge.[41]

3.2   Species Diversity

Long-term data on species diversity at specific locations or habitats are generally lacking but, for example, Welch and Scott[42] report 20-year trends in the plant species richness and composition of 15 moorland sites in Scotland. Such time series can be used to detect fluctuations in species diversity but comparisons between two periods can also reveal long-term trends; Linusson *et al.*,[43] for example, used two datasets, one from the 1960s and one from 1990, to show changes in the species composition of semi-natural grasslands in Småland, southern Sweden. More widespread trends are being quantified through initiatives such as the UK's Environmental Change Network[44] and Countryside Surveys.[45]

3.3   Species Abundance and Biomass

For plants, monitoring usually includes assessments of cover and/or biomass.[42,46] The periodic assessments of the UK Countryside Surveys have revealed different trends in plant diversity in different habitats: although plant diversity between 1990 and 1998 increased in arable field boundaries, it decreased in agriculturally improved grasslands, road verges and streamside vegetation.[37] Using archive biological information, McCollin, Moore and Sparks[47] demonstrated changes in the commonness of plant species between the 1930s and 1990s in different habitats in Northamptonshire in England.

For mobile species such as insects and birds, data on trends in abundance are frequently available. Good data now exist on long-term trends in the abundance and diversity of butterflies, moths and some other insects. Trends in the abundance of mirids and other Heteroptera, for example, have been obtained from a single light trap for over 67 years.[48] Light traps have also been used to quantify long-term trends in single species of moths[49] and to analyse general trends in macrolepidoptera.[50] Data on British macrolepidoptera have been collected by the Rothamsted Insect Survey since 1968. In analysing the data collected over 35 years, 54% of the 338 species investigated had undergone a significant decline in abundance and 22% had shown a significant increase.[51] Long-term data on the abundance of British butterflies have been provided by the Butterfly Monitoring Scheme.[52]

Although there is some information on trends in insect species across Europe, notably on butterflies,[53] the geographical extent of the monitoring networks for insects is not as good as that for birds. Information on the abundance of birds has been collected from many years[7,54] and reported first in the UK as a headline indicator of biodiversity[55] and now across Europe in the Wild Bird Indicator derived from annual breeding bird surveys in 18 European countries, obtained through the Pan-European Common Bird Monitoring Scheme.[56]

Information on the abundance, biomass, average size and/or trophic structure is available for many harvested species. The most notable example of this is fish.[57] Amongst terrestrial species, data exist for some game species.[58] Data on wetland birds, many of which are hunted, are extensively collected.[59]

3.4 Distribution of Species

Information on the distribution of individual species is available for many groups of organisms. In the UK, the BRC has produced atlases for butterflies, plants, fish and other taxa.[8,19,20] Information on changes in distribution is less available but Thomas *et al.*[5] analysed data on changes in the distribution of plants, birds and butterflies in Britain over the last 20 to 40 years and demonstrated declining distributions for all three taxa, particularly the butterflies, which disappeared on average from 13% of the 10 km squares between the 1970s and the 1990s.

3.5 Threatened Status of Species

Data on threatened species, principally the data collated by the IUCN, provide another source of information on biodiversity (see above). In Europe, for example, 12% of the 576 butterfly species known to occur are regarded as threatened.[60] Recently a measure of trends in threatened species has been developed by the IUCN-SSC Red List Programme. In Europe, this Red List Index is based on information on European Red List species and (other) species listed in the annexes of the Birds and Habitats Directives.[61]

Taken together, these sources of information on biodiversity will provide an improving assessment of trends in biodiversity in Europe. As discussed in the previous section, there are many gaps in our knowledge of biodiversity across Europe: these gaps are particularly serious in relation to quantifying trends in most components of biodiversity and even in better-known components such as birds. This problem is made worse by the fact that Europe's biodiversity has changed so much in the past and is likely to face continued pressure. These aspects are considered in the following sections.

4   Biodiversity in Europe: an Historical Perspective

A historical review of European biodiversity places the present situation in a temporal perspective and emphasises the balance between cultural and natural processes in the generation and maintenance of characteristic European biodiversity. Two important conclusions can be drawn from a historical survey: (1) natural changes to European biodiversity occur on a continual basis in response to climate change, so the concepts of equilibrium and long-term stability are not useful in this context; (2) natural and anthropogenic drivers and pressures have developed simultaneously in shaping European biodiversity. The anthropogenic drivers are generally more pervasive, rapid and extreme, but these two sets of drivers are so closely linked that the concept of a natural baseline is of only theoretical interest.

Europe's recent geological past (last 2 million years) was characterised by extreme and rapid climate change. These climate changes forced the species to move, a process that was influenced by physical geography and the location of mountain chains and seas. One major consequence has been that loss of diversity has exceeded creation of diversity by evolutionary processes and Europe is relatively species poor compared with comparable regions in Asia and America[62] (Figure 1).

There is a great contrast between the distribution of biomes during the last glacial maximum (*c.* 22 000–14 000 years ago) and the present day. Forest species in particular had very restricted distributions for long periods of time that inevitably led to loss of species and genetic diversity. As forest area increased in response to altering climate, species characteristic of open habitats experienced habitat fragmentation and loss. These long-term, natural dynamics of habitats comprise an important background to understanding present-day biodiversity status and trends. A significant consequence of repeated glaciations is the concentration of genetic diversity in glacial refuge areas on the margins of Europe. Refugia are usually in the mountains with large topographic variation, offering a variety of contrasting habitats within a small area. Recent genetic analyses have confirmed the importance of refugial regions as current centres of long-term diversity, making such regions of particular importance for biodiversity protection.[63] European biodiversity development during the last 10 000 years has been driven by a combination of natural, climate-driven spreading of species from glacial refugia interacting with increasing anthropogenic influence. A general correspondence between European and North American tree spreading since the last ice age has been used to

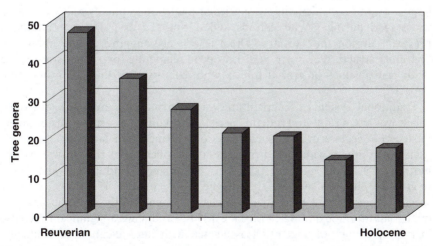

Figure 1   The number of woody genera recorded as fossils from selected inter-glacials during the last two million years in NW Europe.

suggest that climatic change and spreading dynamics from glacial refugia were the major determinants of post-glacial range changes in forest trees.[64] A close correspondence between present-day observed and modelled tree species distributions in Europe, where climatic variables interacted with physiological constraints to generate the modelled species range limits also indicates the importance of climate in determining species ranges[65,66] (Figure 2).

The spread of settled agriculture in Europe is a further key influence on European biodiversity. Neolithic agricultural methods spread from south-east to north-west Europe between 9000 and 5000 years ago.[67] Its major consequences were alteration of the balance between forest and open habitats, massive population increases in species directly (crops, domestic animals) and indirectly (weeds, pests) associated with agriculture. Its impact is recorded in pollen diagrams throughout Europe, typically as increases in the number of common species present on the landscape.

Archaeology and palaeoecology have documented the long history of agricultural development within Europe. It is this history that was the foundation for the biodiversity that has been impacted by industrialisation and intensification of land use that forms the background to the current biodiversity crisis. Studies of the past record an evolution of tools and techniques used to alter land surfaces that have impacted biodiversity. A study from southern Sweden documented systems of shifting cultivation (3000 BC–1000 BC) that utilised stone, bone, then wooden tools, fire, domesticated plants and animals. These systems developed into settled agricultural systems (1000 BC–0) that were based upon use of metal tools, hay meadows and manuring.[68] The palaeoecological record indicates that these earlier systems of agriculture led to increased local diversity of higher plants, but decreased diversity associated with forest systems. Doubtless some forest-dependent insects or decomposers were lost or had their populations seriously reduced, but the widespread opening up of forest structure and the increased proportion of open land benefited a very large group of light-demanding species of plants and animals that were more widespread during glacial periods and during the dry, warmer conditions prevalent during the Tertiary. There is a lively ongoing debate about the "natural" structure of north-west European temperate forest that stresses the biodiversity values of grazed forests with extensive but light anthropogenic impact.[69]

Traditional systems of agriculture varied in intensity in space and time, which was favourable for ruderal species that require occasional disturbance. Palaeoecological records document periods of increased landscape exploitation; for example, widespread pressure at the opening of the Bronze Age in north-west Europe, but also periods of landscape abandonment and forest re-growth, such as during the plague years of the 1300s. Patterns of species dynamics and local diversity changed significantly with the onset of industrialisation beginning during the 1800s, producing a significant increase in rate of change of vegetation associated with intensification of landscape management.

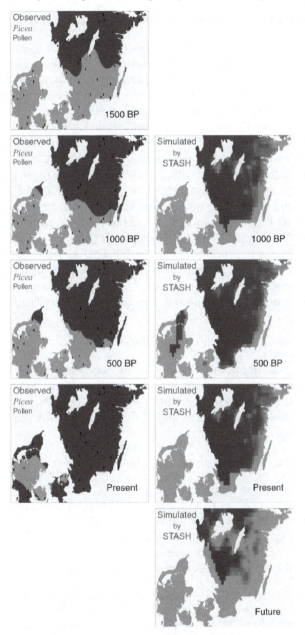

Figure 2 Observed and simulated *Picea* distributions during the last 1500 years. The observed distributions are reconstructed from fossil pollen data. The simulated distributions are generated by the bioclimatic model STASH. The predicted future distribution assumes an atmospheric CO_2 concentration twice that of present.

5 Biodiversity in Europe: Current and Future Threats

The biodiversity that we now have in Europe is threatened by many pressures, particularly habitat degradation, fragmentation and loss. These pressures, often considered together as land-use change, are discussed below. Other pressures include pollution, particularly eutrophication and nitrogen deposition, climate change, invasive organisms, and hunting and fishing pressure.[6,70,71] Some of the pressures on biodiversity are generic but many are specific to the main landscape types occurring in Europe: agriculture, grasslands, forests, wetlands and aquatic habitats, and uplands.[72,73]

Agricultural landscapes, including arable land, permanent grassland and permanent culture, cover about 43% of the European Union. Modern European agriculture is characterised by low labour input, large fields, a loss of genetic diversity both for crops and animal production, high yielding varieties, heavy applications of fertilisers and biocides and removal of landscape features such as woodlands and hedges. Many endangered and threatened species in Europe are now those that depend on the gradually diminishing areas where traditional forms of agriculture still exist.[74,75] Three main categories of pressure threaten biodiversity in agricultural areas, namely intensification, abandonment and scale of operation:[76,77]

1. Intensification of agriculture, particularly:

   - fertiliser application leading to eutrophication,
   - application of biocides, antibiotics and endocrine substances,
   - release of genetically modified organisms (GMOs),
   - conversion of extensive land use into high-intensity production areas,
   - improved field drainage.

2. Abandonment of agriculture, leading to:

   - afforestation though planting,
   - expansion of scrub and woodland through natural regeneration,
   - conversion into urban areas,
   - loss of old varieties of cultivated species,
   - loss of the cultural heritage associated with ancient agricultural landscapes and types of production systems.

3. Change in scale and organisation of agriculture, particularly:

   - monoculture, often a consequence of simplified crop rotations,
   - loss of small-scale heterogeneity in landscape features,
   - loss or disruption of dispersal routes for plants and animals.

The relative importance of these three categories differs among the biogeographical regions of Europe. In Eastern Europe, for example, abandonment of traditional forms of agriculture is a major current pressure on biodiversity.

Pressures on biodiversity are continually changing. Genetically modified organisms are often considered to be a major threat to biodiversity but there is little evidence of a direct impact.[78,79] The main threat of GMOs is that they support further intensification of agriculture, particularly the use of herbicides.[80] This may have consequences for biodiversity similar in extent to the wide-reaching consequences that followed previous changes in agricultural practices such as the development of agro-chemicals.[81,82]

Although technological development is a major driver of the pressures which directly threaten biodiversity, there are others, notably the Common Agricultural Policy (CAP), the establishment of the Trans-European traffic networks (TENs), large-scale demographic and socio-economic changes, and landscape-related policy and planning mechanisms at the national level.[76]

Loss of forest habitats has occurred in Europe for centuries. Europe (including the Russian Federation) covers 2260 million ha, of which 1007 million ha are natural forest and 32 million ha are plantations.[83] Although deforestation is a global problem, particularly in the tropics, the major pressures on forest biodiversity in Europe relate to changing demands on its forests.[84,85] In the eighteenth century, the demands placed on forests in Europe escalated due to population growth. Concerns about wood shortage triggered systematic forest management in Europe to increase productivity, control the rate and type of exploitation and conserve the area of forest.[86] This approach was adopted across large parts of Europe, particularly in landlocked countries that did not have access to the sea and hence could not readily import wood from other parts of the world. This advent of silviculture and the intensification of management gave rise to a range of pressures on biodiversity,[84,85] comprising:

1. Overall changes in forest management:

   - changes in ownership structure, *e.g.* concentration of ownership, commercialisation of state forests,
   - changes in systems for transportation of wood to industry, *e.g.* to road transport,
   - changing of planning strategy, *e.g.* regional focusing of timber harvesting,
   - suppression of natural forest fires in naturally fire-prone forest types.

2. Changes in silvicultural systems:

   - changes in harvesting, *e.g.* introducing clear cutting, and/or increase in size of areas cut,
   - shortening of crop rotation length,
   - introduction of exotic species and plantation forestry,
   - installation and/or alteration of drainage systems,
   - use of fertilisers, pesticides and herbicides,
   - removal of dead wood and diseased trees.

3. New technologies:

- new machinery for timber harvesting and, for example, treatment of regeneration areas (*e.g.* soil scarification),
- new types of forest roads.

Globally, the biodiversity of freshwater systems is rapidly deteriorating as a result of human activities.[87] Across Europe, a number of pressures act singly or in combination in deleteriously affecting the biodiversity of freshwater ecosystems. The type and severity of pressures affecting freshwater ecosystems differ across Europe. For example, although there is no shortage of water for Europe as a whole, regional imbalances can occur between supply and demand. In southern Europe, water availability is often considered as a severe problem and the combined effect of overexploitation and drought have markedly affected regional and local biodiversity.

The eutrophication of freshwater ecosystems occurs in all parts of Europe where intensive agriculture is prevalent and is a major pan-European problem.[6,88] The effects of nutrient enrichment by phosphorus and nitrogen are well documented, with excessive growths of algae resulting in increased turbidity, shifts in community composition and subsequent oxygen deficiency in bottom waters[87] (Table 2). In some instances population growths of toxin-producing species may also result, strongly affecting the usability of the water body. Phosphorus inputs are predominantly from point sources (domestic and industrial), but with the implementation of water treatment facilities and improved chemicals (*e.g.* low phosphorus detergents), eutrophication is expected to decrease. However, significant effects are evident from non-point sources, particularly as surface runoff from agriculture but also from forestry. Both activities may result in nutrient enrichment of receiving waters and loss of biodiversity. In addition, inputs of inorganic matter (*i.e.* sediment) from both agriculture and forestry activity may result in loss or degradation of habitat due to siltation.

The effects of organic pollution on the biodiversity of freshwater systems is another area of concern, particularly in densely populated areas with no or insufficient treatment of sewage effluents.[87] For example, the effects are most prevalent in many Eastern and Southern European countries where advanced sewage treatment is lacking or rare. The effects of organic pollution on aquatic life are similar to those outlined above for eutrophication – high levels of organic input result in major shifts in community composition, ultimately resulting in communities predominantly consisting of only a few tolerant species. However, in contrast to the input of inorganic nutrients and subsequent changes in trophic structure, sewage inputs often are also laden with a wide variety of other pollutants (such as pathogens and heavy metals) that may affect aquatic life.

A third major pressure affecting aquatic ecosystems in large areas of Europe is the acidifying effect of sulfur and nitrogen.[87] Acidification has had profound effects on the biodiversity of lakes and streams, in particular systems situated in

Table 2   Examples of common pressures and related impacts that may lead to the loss of biodiversity in freshwater ecosystems in Europe.

Activity and/or pressure	*Impacts*
Agriculture and Forestry	
Fertiliser use and ploughing	Erosion, increased nutrient load leading to eutrophication
Application of agrochemicals and antibiotics	Contamination with toxic or ecotoxic compounds leading to alteration of population dynamics and community structures
	Scarification
Forestry and afforestation	Enhanced water runoff, with consequences for erosion, eutrophication and local acidification
Alterations of natural hydrology	
Irrigation, drainage	Change of natural hydrology leading to habitat
Water abstraction	alteration and destruction (such as loss of
Flood prevention (damming)	wetlands) migration barriers
Channelisation	
Acidification	
Acidifying effects of runoff	Alteration of community composition
Metal (Al) toxicity	
Waste disposal (organic pollution)	
Sewage, treated wastewater	Contamination with nutrients, organic
Slurry	compounds, toxic or ecotoxic compounds
Direct disposal of waste	leading to eutrophication, alteration of population dynamics and community composition
Fishing, fishery and pisciculture	
Fish farming: use of antibiotics, direct feeding or fertilisation	Same impacts as waste disposal Alteration of community composition
Fish stocking and harvest of stocked fish	Dilution of native genetic stocks
Harvest of natural stocks	
Introduction of new species	
Exotic fish	Alteration of the community composition, leading
Other species	to displacement of indigenous species

the southern parts of Norway, Sweden and Finland. To mitigate the deleterious effects of acidic deposition, liming activities are widespread (particularly in Sweden). In the last decade S and N emissions have decreased markedly and recent studies have shown "chemical" recovery of some ecosystems.[89] However, in many areas and ecosystems, biological recovery will greatly lag behind that of chemical recovery due to the slow processes of recolonisation by many aquatic organism groups (*e.g.* fish and invertebrate groups that lack aerial dispersal mechanisms).

Although threats to biodiversity can be considered ecosystem-by-ecosystem, policies and actions in one ecosystem can influence other ecosystems. For example, policies pursued at local, regional or national scales may favour economic development and urbanisation, agriculture or grazing, all of which may have an impact on, for example, forest biodiversity.[84] In addition, changing land ownership patterns and economy, such as those taking place in the eastern parts of Europe, may lead to conflicts with biodiversity conservation. A range of policies may also indirectly result in land-use changes. For instance, depopulation of rural areas leads to abandonment of land, including forested land. This may lead to both positive and negative outcomes for biodiversity. One negative outcome is an increased risk of forest fires. In southern Europe (Spain, France, Greece, Italy, Portugal), the area burnt increased exponentially between 1970 and 2000,[90] with an average of almost 480 000 hectares annually destroyed by fire.

A further important consideration is that human-generated pressures can cover large spatial scales, affecting regions often far away from the area of origin. For example, the biodiversity of lakes and streams in the Nordic countries, particularly the southernmost regions of Norway, Sweden and Finland, has been altered by the deposition of sulfur and nitrogen (SO_4 and NO_x) compounds emanating from elsewhere.[87]

6   The Policy Response to Biodiversity Loss

Action to prevent biodiversity loss began long before the CBD was born in 1992. The French Nature Conservation Act, for example, was established in 1976, the UK Access to the Countryside and National Parks Act in 1949 and the establishment of protected areas started in the nineteenth century.[2] In Europe, the first Environmental Action Programme was launched in 1973, the Birds Directive in 1979 and the Habitats Directive in 1992. However, the CBD led to a rapid increase in the development of policy on biodiversity in Europe and elsewhere. At the national scale, biodiversity strategies have been published in Austria (1998), Denmark (1995), Finland (1997), Ireland (2002), The Netherlands (1995, revised 2002), Portugal (2001), Spain (1998), Sweden (1994), UK (1994), Estonia (1999), Latvia (2000), Lithuania (1996), Poland (1997), Slovenia (2001) and Slovakia (1997) and are in preparation in other countries. The European Community published a biodiversity strategy in 1998, based on a policy of incorporating biodiversity concerns in sectoral policies[91] and adopted four Biodiversity Action Plans in 2001 – on the Conservation of Natural Resources, Agriculture and Fisheries and on Economic and Development Cooperation. In 2006, the European Commission published a policy communication on biodiversity.[92] At the pan-European level, the Pan-European Biological and Landscape Diversity Strategy (PEBLDS) was established to stop and reverse the loss of biological and landscape diversity in Europe.

In Europe the flagship initiative on biodiversity is Natura 2000, a network of protected sites containing natural habitats of the highest value and species that are rare, endangered or vulnerable in the European Community. The Natura 2000 network includes Special Areas of Conservation (SACs) designated under the Habitats Directive, which support natural habitats and species of plants or animals other than birds, and Special Protection Areas (SPAs) classified under the Birds Directive, which support wild birds and their habitats. The Natura 2000 network has met with mixed success. The UK has had a long history of implementing legislation for nature conservation, including recent amendments to the 1981 Wildlife and Countryside Act. The main focus of this legislation has been the notification, protection and management of a national series of Sites of Special Scientific Interest. The many landowners involved (approximately 25 000) have long been used to the issue of statutory designations and so the relatively recent proposals for Natura 2000 comes on top of an already well-established system of both regulation and financial incentives. In other parts of Europe, however, there has been strong local opposition to designation of Natura 2000 sites.[93,94]

Despite the progress being made in combating biodiversity loss in many species and habitats following the implementation of the CBD across Europe, there are few signs that the rate of loss is declining. Indeed, a recent report concluded that "The living world is disappearing before our eyes".[95] However, as argued above, the available information on trends in biodiversity covers only a small fraction of Europe's habitats and species and is an inadequate basis for policy makers seeking to halt biodiversity loss.[96]

7 Quantifying Biodiversity Loss

The fundamental problem with quantifying biodiversity loss is the enormous variety of habitats and species: there will always be insufficient resources to monitor every European habitat and species. Even the monitoring of Natura 2000 sites represents an enormous challenge: as of March 2005 there were 19 516 candidate SACs and 4169 SPAs.

In response to this problem, biodiversity indicators have been proposed as means of providing rapid measures of biodiversity by both researchers[32,97] and policy makers.[96] From the policy perspective, the CBD has led the recent development of biodiversity indicators: eight biodiversity "focal" indicators were considered ready for immediate testing while another 13 were recommended for further development (Decision VII/30 (CBD COP7); see Table 3). This Decision notes the need for a framework to:

- facilitate the assessment of progress towards the (CBD) target to significantly reduce the rate of loss of biodiversity by 2010 and the communication of this assessment;
- promote coherence among the programmes of work of the CBD;
- provide a flexible framework within which national and regional targets may be set, and indicators identified.

For the EU, a set of European biodiversity headline indicators was adopted at the Malahide stakeholder conference "Biodiversity and the EU: Sustaining life, sustaining livelihoods" in May 2004. These indicators were subsequently taken up in the Council Conclusions (Environment, June 2004). This list shows close

Table 3   Summary of CBD focal and EU biodiversity headline indicators.

EU biodiversity headline indicators [with variation from CBD focal indicator noted where appropriate][a]
Trends in extent of selected biomes, ecosystems and habitats
Trends in abundance and distribution of selected species
Coverage of protected areas
Change in status of threatened and/or protected species [CBD focal indicator does not include protected species]
Trends in genetic diversity of domesticated animals, cultivated plants and fish species of major socioeconomic importance
Area of forest, agricultural, fishery and aquaculture ecosystems under sustainable management [CBD focal indicator does not include fisheries]
Nitrogen deposition
Numbers and cost of alien invasions
Impact of climate change on biodiversity [Not included in CBD list]
Marine trophic index
Water quality in aquatic ecosystems [CBD focal indicator refers only to freshwater ecosystems]
Connectivity/fragmentation of ecosystems
Funding to biodiversity [Not included in CBD list]
Public awareness and participation [Not included in CBD list]
Ecological footprint and related concepts[b]

[a]CBD list also includes: status and trends of linguistic diversity and numbers of speakers of indigenous languages; official development assistance provided in support of the Convention.
[b] Additional indicator added in 2006.

overlap with the CBD list (Table 3). An overlapping set of pan-European biodiversity indicators has also been developed through PEBLDS. The Streamlining European 2010 Biodiversity Indicators initiative (SEBI2010) has been established by the European Environment Agency (EEA) to coordinate the implementation of these indicators.

These indicators vary markedly in the degree to which they are ready for use.[96] For example, a strong candidate for the implementation of the focal indicator "Trends in abundance and distribution of selected species" is the European Wild Bird Index, based on the UK Wild Bird Index[98] applied across Europe,[56,99] and the Red List Index[61] provides a measure of "Change in status of threatened and/or protected species". These indicators could probably be rapidly implemented.[96] Three other indicators are close to implementation: the marine trophic index,[57,100] coverage of protected areas and trends in habitat extent.[96]

Although the European Wild Bird Index and the Red List Index already have the potential to provide valuable information on trends in many European species, the degree to which these and other indicators provide adequate information on general trends in biodiversity remains unclear. The main concern is that trends in one group of species do not reflect trends in other groups. A number of studies have critically evaluated this concern and these have usually shown that spatial and temporal trends in the diversity of one taxon provide poor predictions of the diversity of other taxa.[101–109] For example, the species richness of birds and seven invertebrate taxa were found not to correlate with each other along a gradient of forest disturbance in Cameroon.[103]

These studies usually considered the congruence of different taxa in either different habitats in relatively small areas[103] or in the same habitat types in different parts of the same country.[109] The purpose of the BioAssess project, involving partners from ten European countries and conducted between 2000 and 2004, therefore, was to assess the potential of a range of taxa as indicators of biodiversity in managed terrestrial habitats in Europe. Using standardised protocols, the diversity of different taxa was sampled in 64 sites across Europe (see below). The following taxa were chosen both as representing major components of biodiversity and as potential indicators of biodiversity; birds, butterflies, lichens, plants, ground beetles (Carabidae), soil macrofauna (invertebrates greater than 1 cm in length) and soil Collembola.

No single taxon was found to be an adequate indicator of the diversity, as measured by species richness, of all other taxa. There were, however, significant relationships between the species richness of several taxa, notably between plants, lichens, butterflies and birds. In contrast, no significant relationships were found between the species richness of the three other taxa studied, two soil-dwelling taxa, soil macrofauna and soil Collembola, and epigeal (ground-surface dwelling) ground beetles. Furthermore, the species richness of above-ground taxa generally failed to correlate with below-ground biodiversity. This study suggested that some species-based indicators, such as birds, might succeed in predicting the status and trends in some other major components of biodiversity. International initiatives to monitor trends in the abundance and diversity of birds and associated indicators such as the European Wild Bird Index[56] may therefore be important in indicating trends in some other components of biodiversity. However, this study also showed that this index would fail to predict trends in the diversity of many taxonomic groups, particularly soil-dwelling and epigeal invertebrates.

8 Biodiversity and Land-use Change

Although monitoring trends in biodiversity is a major challenge, information on trends alone, without an assessment of what is causing these trends, provides an inadequate basis for taking action against biodiversity loss. We know, as discussed above, that there are many anthropogenic drivers of biodiversity loss. However, we have very little quantitative information on the contribution of each of these drivers. There are some notable exceptions. Our understanding of the impacts on biodiversity of aquatic pollutants[87] and agricultural practice[81] is relatively good. This understanding has led to specific measures to alleviate these impacts, although some of these have not been as effective as intended.[110]

Although studies on particular species and habitats are increasing our understanding of the impact of many of the pressures affecting biodiversity, there have been few assessments of the impact of major drivers across Europe. The MIRABEL framework, however, was developed to assess the impact of land abandonment, eutrophication, afforestation and other pressures on bio-diversity in 13 European ecological regions, and showed that agricultural intensification is one of the main threats to European biodiversity.[70]

General assessments of the impact of different drivers on biodiversity are made difficult by possible differences in the impact of these drivers on different species and habitats. Research in forest ecosystems, for example, demonstrates that different invertebrate taxa respond in different ways to forest manage-ment.[111,112]

In an attempt to provide an assessment of the impact of major land-use changes across Europe, the BioAssess project (see above) measured the diver-sity of several taxa along gradients of land-use intensity in eight countries (Portugal, Spain, France, Switzerland, Hungary, Ireland, Finland and the UK), thereby encompassing six major biogeographical zones (Boreal, Atlantic, Continental, Alpine, Mediterranean and Pannonian). A total of more than 3450 species were recorded from six $1 \, \text{km}^2$ sites in each of eight countries, ranging from 1442 plant to 111 bird species. Mean species richness across the land-use gradient ranged from 321 plant to 34 butterfly species. In terms of species richness, different taxa varied in their sensitivity to land use: the impact of land use ranged from a mean difference of 40% (27 species) between the species richness of carabids recorded in the most and least species-rich sites to a mean difference of 25% (17 species) between the species richness of soil Collembola. The most speciose taxa, plants and lichens, showed a 31% (98 species) and 35% (61 species) difference, respectively. Several taxa responded similarly to land use: plants, birds, butterflies and lichens showed similar trends in species richness across the land use gradient (Figure 3). The diversity of each of these taxa was greatest in sites with a mixture of land uses and tended to be lowest in managed forests and sites dominated by arable fields. The soil-dwelling and epigeal invertebrates, however, responded differently. Carabids, for example, showed an increase in diversity across the gradient from forest to agricultural sites.

The BioAssess project confirms that biodiversity is unlikely to show a simple, unified response to major anthropogenic drivers such land-use change: these drivers may have different impacts on different groups of plants and animals. Indeed, individual species within each taxa are also likely to respond differently, as analyses of the BioAssess data are beginning to dem-onstrate.[113-115]

9   Discussion

Despite our gaps in knowledge and the lack of data, the evidence of biodiversity loss in Europe is convincing.[5,116,117] Further pressure on land use is likely to have a continued impact on biodiversity, as is climate change[118] and the interaction between climate change and land-use change.[24,119]

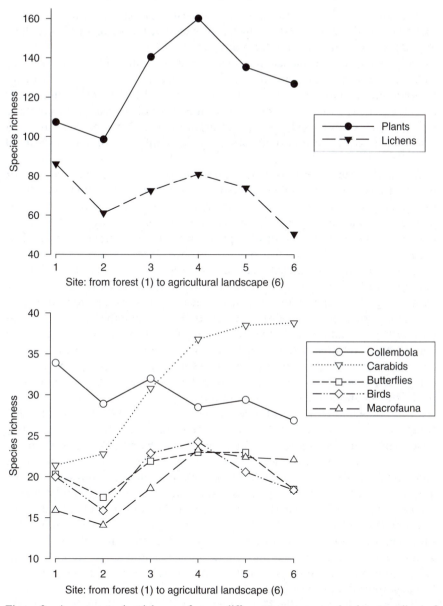

Figure 3   Average species richness of seven different taxa across a land-use gradient of sites from unmanaged forest to intensive arable in eight different countries in Europe.

The policy responses to biodiversity loss in the last twenty years have demonstrated a commitment to conserve biodiversity but despite major initiatives, notably NATURA 2000 and the growing list of Biodiversity Action Plans, there are worrying signs that action to conserve biodiversity is not always effective.[110,120] And despite the increasing amount of available information on biodiversity, most of which is provided by the action of NGOs, there are inadequate data available for many taxa.

The implementation of headline biodiversity indicators in Europe is to be welcomed but it is unlikely that their implementation will provide direct information on many taxa other than birds. The Red List Index[61] will provide an assessment of a range of taxa but it is biased towards species groups that are well known. As discussed above, indicators derived from information on a restricted range of species do not provide reliable information on trends in all taxa.

We clearly need better information on trends in biodiversity, particularly for taxa that have been relatively neglected. We also need information that allows us to quantify the impact of major drivers of biodiversity and to evaluate the policies and measures put in place to conserve biodiversity loss. But it is unrealistic to expect that monitoring programmes to assess trends in more than a very few taxa will be established across Europe, at least at the scale that bird monitoring is being conducted. We therefore argue for a two-level approach to monitoring biodiversity in Europe and elsewhere, the first comprising extensive monitoring of a limited number of taxa, such as birds, which are both intrinsically important and are also potential indicators of other components of biodiversity. In addition to the monitoring initiatives on birds referred to above, there are other extensive monitoring programmes in various stages of development.[121] Most of these programmes have a limited geographical coverage but methods do exist to aggregate data from a variety of sources to provide comparable indicators of biodiversity.[122]

Given the limited value of one or few taxa as indicators of all other components of biodiversity, however, a network of a limited but carefully selected number of sites is needed for the intensive observation of a wider range of taxa. Such a network should provide information on those components of biodiversity that extensive monitoring programmes do not address. It should also serve as a focus for research, evaluating indicators of biodiversity used in extensive programmes and developing new indicators, including indicators based on earth observation. In addition, this network of intensively monitored sites would provide a basis for an improved understanding – through observation, experimentation and modelling – of the impact of anthropogenic drivers and pressures on biodiversity in the context of natural temporal and biogeographical influences on biodiversity.

To address effectively the 2010 targets to reduce or halt the loss of biodiversity we need information on trends on biodiversity delivered to those that need it to take the required action in the timescale needed to do so.[123] Effective action does, however, require that we have a much better understanding of the impacts of various anthropogenic drivers and pressures on biodiversity. A network of intensively monitored observation and research sites, linked to the extensive monitoring networks of a limited number of taxa such as birds, would therefore serve to provide both an improved understanding of trends in biodiversity and also a better understanding of their causes. It would also serve to bring scientists, policy makers and other stakeholders closer together and, therefore, improve the likelihood of effective action to conserve biodiversity.

References

1. E. O. Wilson, *BioScience*, 1985, **35**, 700.
2. B. Groombridge (ed.), "Global Biodiversity: Status of the Earth's Living Resources", Chapman and Hall, London, 1992.
3. V. H. Heywood (ed.), "Global Biodiversity Assessment", UNEP/University of Cambridge, Cambridge, 1995.
4. R. Dirzo and P. H. Raven, *Annu. Rev. Environ. Resour.*, 2003, **28**, 137.
5. J. A. Thomas, M. G. Telfer, D. B. Roy, C. D. Preston, J. J. D. Greenwood, J. Asher, R. Fox, R. T. Clarke and J. H. Lawton, *Science*, 2004, **303**, 1879.
6. D. Stanners and P. Bourdeau, "Europe's Environment – The Dobris Assessment", European Environment Agency, Copenhagen, 1995.
7. BirdLife International, "Birds in Europe: population estimates, trends and conservation status", BirdLife International, Wageningen, The Netherlands, 2004.
8. C. E. Davies, J. Shelley, P. T. Harding, I. F. G. McLean, R. Gardiner and G. Peirson (eds), "Freshwater fishes in Britain – the species and their distribution", Harley Books, Colchester, 2004.
9. Species2000, http://www.sp2000.org.
10. Fauna Europea, http://www.faunaeur.org/.
11. Global Biodiversity Information Facility (GBIF), http://www.gbif.org/.
12. European Network for Biodiversity Information (ENBI), http://www.enbi.info.
13. European Invertebrate Survey, http://www.eis-international.org/.
14. Swedish Species Information Centre, http://www.artdata.slu.se.
15. Luomus project, http://www.luomus.fi/.
16. Swiss Biodiversity Forum, http://www.biodiversity.ch/information/.
17. F. Horváth, Z. Korsós, E. Kovácsné Láng and I. Matskási, "National Biodiversity Monitoring System", Hungarian Natural History Museum, Budapest, 1997.
18. Biological Records Centre (BRC), http://www.brc.ac.uk.
19. J. Asher, M. Warren, R. Fox, P. Harding, G. Jeffcoate and S. Jeffcoate, "The millennium atlas of butterflies in Britain and Ireland", Oxford University Press, Oxford, 2001.
20. C. D. Preston, D. A. Pearman and T. D. Dines (eds), "New Atlas of the British and Irish Flora", Oxford University Press, Oxford, 2002.
21. National Biodiversity Network (NBN) Gateway, http://www.search-nbn.net.
22. S. Reddy and L. M. Davalos, *J. Biogeogr.*, 2003, **30**, 1719.
23. P. F. Donald, R. E. Green and M. F. Heath, *Proc. Roy. Soc. Lond. B Biol. Sci.*, 2001, **268**, 25.

24. M. S. Warren, J. K. Hill, J. A. Thomas, J. Asher, R. Fox, B. Huntley, D. B. Roy, M. G. Telfer, S. Jeffcoate, P. Harding, G. Jeffcoate, S. G. Willis, J. N. Greatorex-Davies, D. Moss and C. D. Thomas, *Nature*, 2001, **414**, 65.

25. IUCN, "2006 IUCN Red List of Threatened Species", IUCN The World Conservation Union, 2006.

26. J. A. Thomas, R. T. Clarke, J. A. Elmes and M. E. Hochberg, in "Population Dynamics in the Genus Maculinea (Lepidoptera: Lycaenidae)", J. P. Dempster and I. F. G. McLean (eds), Chapman & Hall, London, 1998, 261.

27. M. Fischer and J. Stocklin, *Conservat. Biol.*, 1997, **11**, 727.

28. F. Igoe, M. O'Grady, C. Byrne, P. Gargan, W. Roche and J. O'Neill, *Aquat. Conservat. Mar. Freshwat. Ecosyst.*, 2001, **11**, 77.

29. R. L. H. Dennis and T. G. Shreeve, *Biol. Conservat.*, 2003, **110**, 131.

30. S. Van der Veken, K. Verheyen and M. Hermy, *Flora*, 2004, **199**, 516.

31. U. Bohn and R. Neuhäusl, "Karte der natürlichen Vegetation Europas", Bundesamt für Naturschutz, Bonn, 2000.

32. T. -B. Larsson (ed.), *Ecological Bulletins*, 2001, **50**, 1.

33. P. Nowicki, "Environmental benefits from Agriculture: European OECD Countries", OECD, Paris, 1997.

34. H. Pereira, T. Domingos and L. Vicente (eds), "Portugal Millennium Ecosystem Assessment: State of the Assessment", in press.

35. R. Päivinen, M. Lehikoinen, A. Schuck, T. Häame, S. Väätäinen, P. Kennedy and S. Folving, "Combining Earth Observation Data and Forest Statistics", European Forest Institute/Joint Research Centre-European Commission, Joensuu, 2001.

36. J. Puumalainen, P. Kennedy, S. Folving, P. Angelstam, G. Banko, J. Brandt, M. Caldeira, C. Estreguil, J. M. Garcia del Barrio, M. Keller, M. Köhl, M. Marchetti, P. Neville, H. Olsson, J. Parviainen, H. Pretzsch, H. P. Ravn, G. Ståhl, E. Tomppo, J. Uuttera, A. Watt, B. Winkler and T. Wrbka, "Forest Biodiversity – Assessment Approaches for Europe", European Commission. Joint Research Centre, Institute for Environment and Sustainability, Ispra, 2002.

37. R. Haines-Young, C. J. Barr, L. G. Firbank, M. Furse, D. C. Howard, G. McGowan, S. Petit, S. M. Smart and J. W. Watkins, *J. Environ. Manag.*, 2003, **67**, 267.

38. E. M. Bignal and D. I. McCracken, *J. Appl. Ecol.*, 1996, **33**, 413.

39. EEA (European Environment Agency), "High nature value farmland-Characteristics, trends and policy challenges", European Environment Agency, Copenhagen, 2004.

40. EEA (European Environment Agency), "Europe's environment: the third assessment", European Environment Agency, Copenhagen, 2003.

41. R. M. Halahan, "Favourable conservation status – to the heart of EU wildlife legislation", WWF, 2003.

42. D. Welch and D. Scott, *J. Appl. Ecol.*, 1995, **32**, 596.

43. A. C. Linusson, G. A. I. Berlin and E. G. A. Olsson, *Plant Ecol.*, 1998, **136**, 77.
44. Environmental Change Network, http://www.ecn.ac.uk/.
45. Defra, "Countryside Survey 2000 Accounting for Nature: Assessing Habitats in the UK Countryside", UK Department for Environment, Food and Rural Affairs, London, 2002.
46. G. A. I. Berlin, A. C. Linusson and E. G. A. Olsson, *Acta Oecol. – Int. J. Ecol.*, 2000, **21**, 125.
47. D. McCollin, L. Moore and T. Sparks, *Biol. Conservat.*, 2000, **92**, 249.
48. T. R. E. Southwood, P. A. Henderson and I. P. Woiwod, *Eur. J. Entomol.*, 2003, **100**, 557.
49. K. F. Conrad, I. P. Woiwod and J. N. Perry, *Biol. Conservat.*, 2002, **106**, 329.
50. A. D. Watt and I. P. Woiwod, *Oikos*, 1999, **87**, 411.
51. K. F. Conrad, I. P. Woiwod, M. Parsons, R. Fox and M. S. Warren, *J. Insect Conservat.*, 2004, **8**, 119.
52. E. Pollard, D. Moss and T. J. Yates, *J. Appl. Ecol.*, 1995, **32**, 9.
53. C. A. M. Van Swaay, "Trends for butterfly species in Europe", De Vlinderstichting, Wageningen, 2003.
54. W. J. M. Hagemeijer and M. J. Blair, "The EBCC Atlas of European Breeding Birds", T. & A. D. Poyser, London, 1997.
55. M. A. Eaton, D. G. Noble, P. A. Cranswick, N. Carter, S. Wotton, N. Ratcliffe, A. Wilson, G. M. Hilton and R. D. Gregory, "The State of the UK's Birds 2003", RSPB, Sandy, 2004.
56. R. D. Gregory, A. van Strien, P. Vorisek, A. W. G. Meyling, D. G. Noble, R. P. B. Foppen and D. W. Gibbons, *Phil. Trans. Biol. Sci.*, 2005, **360**, 269.
57. D. Pauly and R. Watson, *Phil. Trans. Biol. Sci.*, 2005, **360**, 415.
58. R. Moss and A. Watson, *Adv. Ecol. Res.*, 2001, **32**, 53.
59. M. Frederiksen, R. D. Hearn, C. Mitchell, A. Sigfusson, R. L. Swann and A. D. Fox, *J. Appl. Ecol.*, 2004, **41**, 315.
60. C. van Swaay and M. S. Warren, "Red Data Book of European Butterflies (Rhopalocera)", Council of Europe Publishing, Strasbourg, 1999.
61. S. H. M. Butchart, A. J. Stattersfield, L. A. Bennun, S. M. Shutes, H. R. Akcakaya, J. E. M. Baillie, S. N. Stuart, C. Hilton-Taylor and G. M. Mace, *PLoS Biology*, 2004, **2**, 2294.
62. R. E. Latham and R. E. Ricklefs, in "Continental Comparisons of Temperate-zone Tree Species Diversity", R. E. Ricklefs, D. Schluter (eds), University of Chicago Press, Chicago, 1993, 294.
63. R. J. Petit, I. Aguinagalde, J. L. de Beaulieu, C. Bittkau, S. Brewer, R. Cheddadi, R. Ennos, S. Fineschi, D. Grivet, M. Lascoux, A. Mohanty, G. M. Muller-Starck, B. Demesure-Musch, A. Palme, J. P. Martin, S. Rendell and G. G. Vendramin, *Science*, 2003, **300**, 1563.
64. B. Huntley and T. Webb, *J. Biogeogr.*, 1989, **16**, 5.
65. M. T. Sykes, I. C. Prentice and W. Cramer, *J. Biogeogr.*, 1996, **23**, 203.

66. R. H. W. Bradshaw, B. H. Holmqvist, S. A. Cowling and M. T. Sykes, *Canadian Journal Of Forest Research-Revue Canadienne De Recherche Forestiere*, 2000, **30**, 1992.
67. N. Roberts, "The Holocene", Blackwell Publishing, Oxford, 1998.
68. E. G. A. Olsson in "Agro-ecosystems from Neolithic Time to Present", B. E. Berglund (ed.), Munksgaard, Copenhagen, 1991, 293.
69. F. W. M. Vera, "Grazing ecology and forest history", CAB International, Wallingford, 2000.
70. S. Petit, L. Firbank, B. Wyatt and D. Howard, *AMBIO: A Journal of the Human Environment*, 2001, **30**, 81.
71. J. Young, A. Watt, P. Nowicki, D. Alard, J. Clitherow, K. Henle, R. Johnson, E. Laczko, D. McCracken, S. Matouch, J. Niemelä and C. Richards, *Biodiversity And Conservation*, 2005, **14**, 1641.
72. J. Young, P. Nowicki, D. Alard, K. Henle, R. Johnson, S. Matouch, N. Niemelä and A. D. Watt (eds), "Conflicts between human activities and the conservation of biodiversity in agricultural landscapes, grasslands, forests, wetlands and uplands in Europe", Centre for Ecology and Hydrology, Banchory, 2003.
73. J. Young, L. Halada, T. Kull, A. Kuzniar, U. Tartes, Y. Uzunov and A. D. Watt (eds), "Conflicts between human activities and the conservation of biodiversity in agricultural landscapes, grasslands, forests, wetlands and uplands in the Acceding and Candidate countries (ACC)", Centre for Ecology and Hydrology, Banchory, 2004.
74. G. Kaule, "Arten- und Biotopschutz", Ulmer, Stuttgart, 1991.
75. H. Sukopp, W. Trautman and D. Korneck, *Schriftenreihe für Vegetationskunde*, 1978, **12**, 1.
76. K. Henle in "Agricultural Landscapes", J. Young, P. Nowicki, D. Alard, K. Henle, R. Johnson, S. Matouch, N. Niemelä and A. D. Watt (eds), Centre for Ecology and Hydrology, Banchory, 2003, 25.
77. D. Alard in "Grasslands", J. Young, P. Nowicki, D. Alard, K. Henle, R. Johnson, S. Matouch, N. Niemelä and A. D. Watt (eds), Centre for Ecology and Hydrology, Banchory, 2003, 48.
78. M. D. Hunter, *Agri. Forest Entomol.*, 2000, **2**, 77.
79. M. O'Callaghan, T. R. Glare, E. P. J. Burgess and L. A. Malone, *Annu. Rev. Entomol.*, 2005, **50**, 271.
80. L. G. Firbank and F. Forcella, *Science*, 2000, **289**, 1481.
81. J. R. Krebs, J. D. Wilson, R. B. Bradbury and G. M. Siriwardena, *Nature*, 1999, **400**, 611.
82. A. R. Watkinson, R. P. Freckleton, R. A. Robinson and W. J. Sutherland, *Science*, 2000, **289**, 1554.
83. FAO (Food and Agriculture Organization), "State of the World's Forests 2001", Food and Agriculture Organization of the United Nations, Rome, 2001.
84. N. Niemelä in "Forests", J. Young, P. Nowicki, D. Alard, K. Henle, R. Johnson, S. Matouch, N. Niemelä and A. D. Watt (eds), Centre for Ecology and Hydrology, Banchory, 2003, 68.

85. J. Niemelä, J. Young, D. Alard, M. Askasibar, K. Henle, R. Johnson, M. Kurttila, T. B. Larsson, S. Matouch, P. Nowicki, R. Paiva, L. Portoghesi, R. Smulders, A. Stevenson, U. Tartes and A. Watt, *Forest Pol. Econ.*, 2005, **7**, 877.

86. J. Perlin, "A Forest Journey: The Role of Wood in the Development of Civilization", Harvard University Press, Cambridge, Massachusetts, 1989.

87. R. Johnson in "Wetlands", J. Young, P. Nowicki, D. Alard, K. Henle, R. Johnson, S. Matouch, N. Niemelä and A. D. Watt (eds), Centre for Ecology and Hydrology, Banchory, 2003, 98.

88. V. H. Smith, *Environ. Sci. Pollut. Res.*, 2003, **10**, 126.

89. J. L. Stoddard, D. S. Jeffries, A. Lukewille, T. A. Clair, P. J. Dillon, C. T. Driscoll, M. Forsius, M. Johannessen, J. S. Kahl, J. H. Kellogg, A. Kemp, J. Mannio, D. T. Monteith, P. S. Murdoch, S. Patrick, A. Rebsdorf, B. L. Skjelkvale, M. P. Stainton, T. Traaen, H. van Dam, K. E. Webster, J. Wieting and A. Wilander, *Nature*, 1999, **401**, 575.

90. J. Goldamer, "Proceedings of the XX IUFRO Congress", IUFRO, Kuala Lumpur, 2001.

91. G. Drucker and T. Damarad, "Integrating Biodiversity in Europe. A Review of Convention on Biological Diversity General Measures and Sectoral Policies", European Centre for Nature Conservation, Tilburg, Netherlands, 2000.

92. EC (European Commission), "Halting the loss of biodiversity by 2010 and beyond. Sustaining ecosystem services for human well-being", Commission of the European Communities, Brussels, 2006.

93. S. Stoll-Kleemann, *J. Environ. Psychol.*, 2001, **21**, 369.

94. S. Stoll-Kleemann, *J. Environ. Plann. Manag.*, 2001, **44**, 111.

95. The Royal Society, "Measuring Biodiversity for Conservation", The Royal Society, London, 2003.

96. EASAC (European Academies Science Advisory Council), "A Users' Guide to Biodiversity Indicators", The Royal Society, London, 2005.

97. J. Rainio and J. Niemelä, *Biodiversity and Conservation*, 2003, **12**, 487.

98. R. D. Gregory, D. G. Noble and J. Custance, *Ibis*, 2004, **146**, 1.

99. A. J. Van Strien, J. Pannekoek and D. W. Gibbons, *Bird Study*, 2001, **48**, 200.

100. D. Pauly, V. Christensen, S. Guenette, T. J. Pitcher, U. R. Sumaila, C. J. Walters, R. Watson and D. Zeller, *Nature*, 2002, **418**, 689.

101. J. R. Prendergast, R. M. Quinn, J. H. Lawton, B. C. Eversham and D. W. Gibbons, *Nature*, 1993, **365**, 335.

102. J. R. Prendergast, *Ecography*, 1997, **20**, 210.

103. J. H. Lawton, D. E. Bignell, B. Bolton, G. F. Bloemers, P. Eggleton, P. M. Hammond, M. Hodda, R. D. Holt, T. B. Larsen, N. A. Mawdsley, N. E. Stork, D. S. Srivastava and A. D. Watt, *Nature*, 1998, **391**, 72.

104. P. Duelli and M. K. Obrist, *Agr. Ecosyst. Environ.*, 2003, **98**, 87.

105. I. Oliver, A. J. Beattie and A. York, *Conservat. Biol.*, 1998, **12**, 822.

106. H. R. Negi and M. Gadgil, *Biol. Conservat.*, 2002, **105**, 143.

107. K. Vessby, B. Soderstrom, A. Glimskar and B. Svensson, *Conservat. Biol.*, 2002, **16**, 430.
108. G. A. Krupnick and W. J. Kress, *Biodiversity and Conservation*, 2003, **12**, 2237.
109. N. Sauberer, K. P. Zulka, M. Abensperg-Traun, H. M. Berg, G. Bieringer, N. Milasowszky, D. Moser, C. Plutzar, M. Pollheimer, C. Storch, R. Trostl, H. Zechmeister and G. Grabherr, *Biol. Conservat.*, 2004, **117**, 181.
110. D. Kleijn, F. Berendse, R. Smit and N. Gilissen, *Nature*, 2001, **413**, 723.
111. A. D. Watt, N. E. Stork and B. Bolton, *J. Appl. Ecol.*, 2002, **39**, 18.
112. N. E. Stork, D. S. Srivastava, A. D. Watt and T. B. Larsen, *Biodiversity and Conservation*, 2003, **12**, 387.
113. E. Fedoroff, J. F. Ponge, F. Dubs, F. Fernandez-Gonzalez and P. Lavelle, *Agr. Ecosyst. Environ.*, 2005, **105**, 283.
114. L. T. Waser, S. Stofer, M. Schwarz, M. Küchler, E. Ivits and C. Scheidegger, *Community Ecol.*, 2004, **5**, 121.
115. A. J. Vanbergen, B. A. Woodcock, A. D. Watt and J. Niemelä, *Ecography*, 2005, **28**, 3.
116. D. Maes and H. Van Dyck, *Biol. Conservat.*, 2001, **99**, 263.
117. R. J. Wilson, C. D. Thomas, R. Fox, D. B. Roy and W. E. Kunin, *Nature*, 2004, **432**, 393.
118. C. D. Thomas, A. Cameron, R. E. Green, M. Bakkenes, L. J. Beaumont, Y. C. Collingham, B. F. N. Erasmus, M. F. de Siqueira, A. Grainger, L. Hannah, L. Hughes, B. Huntley, A. S. van Jaarsveld, G. F. Midgley, L. Miles, M. A. Ortega-Huerta, A. T. Peterson, O. L. Phillips and S. E. Williams, *Nature*, 2004, **427**, 145.
119. J. M. J. Travis, *Proc. Roy. Soc. Lond. B Biol. Sci.*, 2003, **270**, 467.
120. D. Kleijn, R. A. Baquero, Y. Clough, M. Diaz, J. De Esteban, F. Fernandez, D. Gabriel, F. Herzog, A. Holzschuh, R. Johl, E. Knop, A. Kruess, E. J. P. Marshall, I. Steffan-Dewenter, T. Tscharntke, J. Verhulst, T. M. West and J. L. Yela, *Ecol. Lett.*, 2006, **9**, 243.
121. Butterfly Conservation Europe, http://www.bc-europe.org.
122. M. de Heer, V. Kapos and B. J. E. ten Brink, *Phil. Trans. Biol. Sci.*, 2005, **360**, 297.
123. R. E. Green, A. Balmford, P. R. Crane, G. M. Mace, J. D. Reynolds and R. K. Turner, *Conservat. Biol.*, 2005, **19**, 56.

Tropical Moist Forests

JON C. LOVETT, ROB MARCHANT, ANDREW R. MARSHALL AND
JANET BARBER

1 Introduction

About 50% of the Earth's surface lies in the tropics between latitudes 30°N and
30°S. This land is in the South American and African continental landmasses
and a scatter of peninsulas and islands in the south Asian and Australian
tropics. More than a third of the world's population inhabit tropical lands and
population growth rates are high. An increasingly high proportion of these
people live in cities, but much of the tropical population relies on subsistence
agriculture. Forest clearance is an important source of land, both through
traditional slash and burn rotations and "frontier" agriculture, where migrat-
ing people are allocated forest land for conversion. Cash crops also play a
significant role in deforestation. Increasingly, tropical agriculture is supplying
markets in industrialised countries. This is resulting in massive transformation
of native tropical forests, usually starting with logging and ultimately leading to
replacement by agriculture. Habitat degradation and loss are the greatest
threats to terrestrial species.[1] Estimates of annual loss of tropical forest range
from 8.7 to 12.5 M ha.[2] An area of between half and equal size to this is
degraded by selective logging each year.[3,4] Loss and degradation of tropical
forests are of global concern as more than half of the world's species are found
in tropical forests, despite covering only 7% of the world's surface.[5] Consequently, the number of species threatened with extinction in tropical forests is
predicted to increase.[6] Tropical forest loss and degradation also have implications for climate change, hydrology, nutrient cycling and natural resource
availability.[7] Restoring degraded forests may therefore be one of the greatest
challenges for ecologists this century.[8]

Conversion and degradation of tropical forest illustrates the fundamental
conflict between conservation and economic development. People in tropical
countries need land for both subsistence livelihoods and cash crops.

Issues in Environmental Science and Technology, No. 25
Biodiversity Under Threat
Edited by RE Hester and RM Harrison
© The Royal Society of Chemistry, 2007

Governments need to develop export agriculture to generate national wealth and this is a central plank for escaping poverty. On the other hand, tropical forest biodiversity is a "common concern of humankind" as defined by the Convention on Biological Diversity and we are obligated to conserve it both for sustainable management for present generations and to meet the needs of future generations. Whilst changes in land cover represent directly observable loss of tropical forests, anthropogenic release of greenhouse gases and consequent shift to a warmer climate change will result in major alteration to the distribution of many species. Global warming highlights the impact of present generations on the future, and it is one of the greatest man-made threats to tropical forest biodiversity.

This chapter first briefly reviews tropical forest ecology and continental-scale patterns of diversity. We then discuss some possible reasons for these broad-scale patterns by looking at plate tectonics, mountain uplift, rainfall and historical climate change. Climate fluctuations over the last 2.2 Myr during the Pleistocene are covered in some detail. This is for two reasons. First, the dramatic climate fluctuations over the last 2 million years, and the last 20 000 years in particular, have left a strong signature in present-day ecology. Second, we need to look to the past to understand what might happen under future conditions of climate change. We also focus on Africa, because this is the continent predicted to be most affected by global warming, so its forests are most under threat. There is also a great deal of local-scale variation in biodiversity. This can be attributed to differences in climate, topography, the biology of individual species and disturbance regime. We then look at past and present anthropogenic impacts on tropical forests, followed by examination of a case study in the mountains of the Eastern Arc tropical forest biodiversity hotspot where cash-crop agriculture is being successfully combined with forest conservation. In conclusion, we review the potential future impacts of global warming and emphasise the need for tropical countries to develop their own research expertise.

2   Tropical Forest Ecology

Tropical forests grow under the climate generated by the inter-tropical convergence zone (ITCZ). The ITCZ is where the trade winds converge in the equatorial low-pressure trough, which is formed at the thermal equator. The thermal equator is a belt of high temperatures caused by solar heating. It migrates north and south in accordance with the relative position of the Earth to the sun. Hot air from the thermal equator rises, condensing as it cools to create tropical rains. The circulation continues as the dry cooler air falls on either side of the ITCZ to create subtropical desert regions. This tropical air-mass circulation is called a Hadley Cell. Movement of the thermal equator creates tropical wet and dry seasons. Closed canopy tropical forests require mean annual rainfalls of more than about 2000 mm without too long a dry season. If rainfall is fairly evenly spread throughout the year then closed-canopy forests can occur at lower rainfalls than this. They also occur on tropical mountains up to an elevation where frost occurs regularly, which is usually around 2400–3000 m. At lower rainfalls, or where there is a long dry season, closed-canopy forest gives way to more open woodland.

If ecology was simple, then patterns of biodiversity would be relatively easy to predict and accordingly straightforward to manage. Theoretical ecologists hypothesise that biodiversity is a function of physical parameters such as soil fertility, rainfall and temperature. So, in a simple world we would expect high-rainfall tropical latitudes to be more biodiverse, with species numbers declining as latitude increases to cooler, more seasonal temperate and boreal lands. To a large extent, this is what we observe. There are more species in the hotter, wetter tropics compared to higher latitudes, but a closer examination of the patterns reveals a bewildering complexity. At a continental scale there are huge differences in tropical forest diversity between South America, Africa and Asia, with Africa being the "odd man out" in that it has much lower species richness. At a regional scale, one range of mountains can host many more species than its neighbours; and at a local scale a diverse forest can be right next to one that is dominated by a few species. It is this complexity that makes the study of tropical forest biodiversity so fascinating. For managers it offers both constraints and opportunities.

Constraints arise because the factors dictating the distribution of tropical species are still largely unknown and often shrouded in historical mystery. It is therefore unlikely that we will ever be sure why species occur where they do. This means that it is very difficult for managers to predict the effects of management practice. For example, will disturbance of a tropical forest by logging have limited long-lasting effects because the ecological determinants of diversity are fixed by temperature and mean annual rainfall? Or will disturbance increase diversity by opening new niches for colonisation by a new set of species? Or will disturbance cause a catastrophic loss of species and transformation of the complex web of life that makes up a mature tropical forest to a simplified ecosystem prone to dramatic changes such as those caused by fire or pest pressure?

Opportunities arise because an extraordinary fact about the spatial distribution of biodiversity over the Earth's surface is that, in terms of numbers of species, it is clustered in a limited number of "hotspots".[9] This discovery opens the possibility of protecting large numbers of tropical forest species by focusing conservation expenditure and activity on the biodiversity hotspots, giving more "bang per buck" of money spent on saving threatened plants and animals. The concept of species having clearly defined patterns of distribution dates back to formulation of floristic kingdoms, with the tropics divided into the neotropics covering South America and the paleaotropics covering Africa and the Indo-Malaysian region, with a separate Australian kingdom.[10] The kingdoms were divided into a series of provinces based on the distribution patterns of the plant species in them. This idea has been extended and refined, most recently by identification of "ecoregions",[11] which are used as a guide to target conservation aimed at alleviating threats to tropical forests. Extent and location of the ecoregions is controversial, so it is interesting to explore the underlying historical dynamics that have led to development of biodiversity hotspots observed today.

3   Continental Scale Variation due to Plate Tectonics

There is a great deal of difference in species diversity between the continents. For example, if we look at distribution in the numbers of species and genera of palms and ferns[12] (Tables 1 and 2) it is clear that continental Africa has comparatively far fewer species of these types of plant than the other tropical areas. Remarkably, even Madagascar, a large island lying off the south-eastern coast of Africa, has more palms and nearly as many ferns as the rest of the continent.

Plate tectonics provides one possible explanation for these patterns, though there are other explanations which we will discuss in later sections. Africa was once at the centre of the super-continent Gondwana, about 180 million years ago, and lay 18 degrees south of its present position so that the equator traversed what is now the Sahara desert.[13] To the north lay the super-continent of Laurasia. In its central position, Africa would have been drier than the western and eastern parts of Gondwana, which were to become South America and Indo-Malaysia, respectively, so it is possible that the extent of wet tropical forest was always less than in the other tropical areas. As Gondwana and Laurasia broke up, North and South America moved westwards from Africa, creating the Atlantic Ocean, joining up via the isthmus of Panama.[14] The huge South American Andean range running along the entire western margin of South America is formed by a tectonic subduction zone, which is still actively uplifting the mountains and creating waves through the Amazon Basin.[14] India

Table 1   Distribution of numbers of species and genera of palms.[12]

Location	Species	Genera
Africa	65	14
South America	550	67
Madagascar	175	16
Indo-Malaysia	1400	100

Table 2   Distribution of numbers of species of ferns.

Location	Species
Africa	650
South America	3500
Madagascar	500
Indo-Malaysia	4500

broke away from eastern Africa to cross what is now the Indian Ocean, crashing into Laurasia to create the Himalayas. Antarctica moved south and Australia and New Guinea moved eastwards to join up with an arc of Laurasian islands that today include Indonesia, Borneo and the Philippines.

High species diversity in Indo-Malaysian forests is thus increased on a regional scale by the area being the meeting point of the Laurasian and Gondwana plates.[16] Two completely different biota have been brought into proximity, effectively doubling the numbers of species. The biological discontinuity in the complex pattern of islands was first noticed by Alfred Russell Wallace and is named Wallace's Line in his honour. A second tectonic reason for high diversity in Indo-Malaysia is the creation of landforms that stimulate the evolutionary process. These include mountain uplift and formation of islands. Both mountains and islands provide new habitats for colonisation and cause genetic isolation, a point noted by Wallace in his early papers. A third reason for the rich biota is high rainfall and humid climate: a topic discussed later.

Following the break up of Gondwana, South America joined with Laurasia causing a wave of plant and faunal immigration from North America, with some South American species also travelling north.[17] The massive Andean uplift created new habitats, stimulating speciation and sending tectonic waves through the Amazon basin to cause fluctuations in river flow and associated ecological dynamics. As with Indo-Malaysia, much of this tectonic activity was in high-rainfall tropical zones.

In contrast, Africa lay in the centre of Gondwana and so was not subject to the major tectonic mountain building activity of Indo-Malaysia and South America. Africa also had a prolonged period of contact with Laurasia and shares many faunal and floral elements. Where mountain building did occur in Africa, through rifting and uplift of the central African plateau, it tended to occur in areas of relatively low rainfall. Where mountains occur under high rainfall, such as the Eastern Arc mountains of Tanzania, the Albertine Rift mountains and mountains in Cameroon and Gabon, they are also rich in species. Indications are that parts of Africa were wetter in the past than at present; for example, Africa was rich in palms in the Cretaceous, though the numbers of species declined about 34 Mya.[18] Plant families such as the Winteraceae and Sarcolaenaceae were present in southern Africa in the Miocene, and are still present on Madagascar, but are now absent from the mainland,[19] though areas of central tropical Africa that are currently dry appear to have been that way for a long period.[20]

Tectonic activity can thus help to explain continental variation in diversity in the tropical forests in three main ways. First, continental drifting can bring together biota that have evolved independently, thereby increasing the numbers of species in a region. Second, plate movement and formation of new islands can create the isolating mechanisms needed for speciation to occur. Third, mountain building can also create new habitats and act as a barrier, stimulating and permitting speciation. Tropical areas that are tectonically active and that are also under high rainfall are exceptionally rich in species.[21] From a management perspective this helps us to locate key areas for biodiversity conservation, and many biodiversity hotspots are in areas where tectonic activity and high rainfall have combined to give high species numbers. However, in terms of threat, these areas are prime places for conversion of forest to agriculture as they have recently developed fertile soils and good rainfall.

4   Regional Scale Variation due to Pleistocene Climate Fluctuations

4.1   Tropical Climate Change

The tendency for climates to change relatively suddenly, even over the past millennia, has been one of the most surprising outcomes of the study of earth history.[22] The current geological period (the Quaternary) is characterised by a series of relatively cool, arid (glacial) phases and relatively warm, humid (interglacial) phases. There have been at least twenty major glacial phases over the course of the Quaternary[23] during which the extent of ice globally was greater than during the intervening interglacials.[24,25] Glacials were also characterised by lower sea levels, differences in the amount of solar radiation reflected by the Earth's surface and changes in atmospheric composition (*e.g.* lower CO_2 content) relative to interglacials. Superimposed upon this major, largely orbitally driven cycle of climate change were numerous lower-magnitude, higher-frequency events. The impacts of these events are recorded over the range of spatial scales, from local to global, while their drivers were often complex feedback mechanisms, such as the interplay between ice sheets and ocean circulation.[26] The only constant regarding climate in the past is that climate has constantly changed, such changes being unevenly felt over the Earth's surface, with certain areas experiencing greater changes in temperature, precipitation and seasonality than others. The maximum extent of ice for the last glacial in other parts of the world may not have coincided with the last glacial maximum (LGM) about 20 000 years ago. For example, there is evidence that the extent of ice on several mountains in eastern Africa reached its maximum in the late glacial, following the LGM, owing to a combination of relatively cool and humid climate conditions.[27] However, the massive ice sheets in the Northern Hemisphere at the LGM will have had a major impact on environmental conditions globally, with world-wide sea levels and monsoon-associated precipitation probably at their lowest.

As more data on environmental change and its ecosystem impacts are produced, a different perspective on the spatial and temporal character of abrupt climate shifts and how these impact on ecosystem composition emerges.[28,29] The tropics, rather than complacently following environmental change recorded at temperate latitudes, are increasingly shown to record changes first,[30] and indeed may act as a pace-setter for change; hence the tropics have hitherto been underestimated in understanding ecosystem response to global climate change.[31,32] Tropical ecosystems may provide an early warning system for climate change, particularly within the present interglacial period when climatic ties to high latitudes have weakened considerably with the demise of the polar ice sheets,[33] a situation that one would expect to continue in the future as ice sheets undergo accelerated contraction. As more long-term ecological data and studies into predicting impacts of climate change on species distribution become available,[34] it is clear that future ecosystem composition, structure and functioning will be different. These parameters respond rapidly to current environmental change and are projected to do so more dramatically in the near future.[35]

4.2 Direct Evidence for Change

One of the foundations for reconstructing past ecosystems is pollen analysis: past vegetation composition and distribution, and changes in this, can be determined by fossil pollen preserved within accumulating sediments whose provenance can be identified back to the parent plant. Assuming that the remains have not been transported far and have been accurately and precisely dated, this type of evidence can be used to gain an insight into the nature of vegetation at a particular time in the past. When these "snap-shot" reconstructions are placed within a time-frame provided by radiocarbon dating, how the vegetation has changed at a single site over time can be reconstructed. Pollen analysis is a remote-sensing tool to enable investigation of long-term ecosystem dynamics;[36] like all remote-sensing tools, there is a need to understand constraints on the spatial resolution attainable. One of the perennial problems for interpreting palaeoecological records is the provenance of the pollen taxa;[37] how reflective of the surrounding vegetation is the pollen accumulating within sediments? This problem of provenance is particularly acute in the tropics where the discipline is relatively new compared to the more-intensively studied temperate latitudes. To identify and quantify provenance a newly established "global" pollen monitoring network[38] will aid in the interpretation of fossil pollen and feed directly into a modelling tool to explore pollen deposition in a landscape scenario. Within the model, floristic elements, landscape characteristics and factors influencing pollen emission, fall speed and climatic factors influencing the pollen deposition can be changed.[39]

A high density of studies has permitted rates and directions of spread of forest taxa to be plotted for Europe and North America.[40] Unfortunately, the availability of the direct evidence required to underpin such studies is the exception rather than the rule, either because conditions conducive to the accumulation of sediments are not present or because of an absence of detailed palaeoecological studies throughout the tropics. There is a large amount of evidence from fossil-based (*i.e.* palaeoecological) studies to indicate that forests do not respond to climate change in a simple, deterministic fashion and as discrete and fixed units. Rather, the evidence suggests that the precise outcome of climate change is far more difficult to predict and is the product of a mixture of the responses of individual taxa, each of which has its own range of ecological tolerances and therefore sensitivity to change. As a result of this, individual behaviour, and because the complex of environmental conditions was unlike those of today, it is highly unlikely that the composition of forests at past time-periods was exactly the same as the present, even in those areas where a forest cover may have persisted since the LGM.

It is clear from the range of palaeoecological archives that the biota in certain locations were more responsive to the climatic vicissitudes of the Late Quaternary than others. Indeed recent interpretations from central Africa[41] and Latin America[42] show that forest ecosystems respond to climate change as a combined individual response of species, resulting in the formation of novel assemblages of taxa. Therefore it is logical that forest cover was present at certain locations at the LGM as intact communities but without modern analogue. In a few cases, however, local edaphic and topographic conditions may have mitigated climate impacts to the extent that forests were able to survive *in situ*, perhaps with relatively minor changes in composition and structure relative to similar forest types today. It is thus difficult to predict the future effect of climate change on tropical forests, as species will respond individualistically and it is likely that individual responses will vary as a function of soil type and ground moisture.

4.3   Inferential Evidence for Change

The nature of past environments, and changes in them, can be determined from indirect sources of evidence. Indirect evidence is mainly in the form of patterns of present-day distributions of species and genetic diversity of forest taxa and associated fauna. It is assumed that the distribution patterns of extant species reflect both past and present-day environmental conditions. Two main patterns of species distributions are commonly referred to as sources of information on past environments. These are levels of diversity, or differences in the number of organisms between areas, and levels of endemism, or differences in the degree of biological uniqueness between areas. Loci of high species diversity and endemism have been used as surrogates for forest refuges[43] under the assumption that high diversity and endemism are facilitated by relative environmental stability through long- and short-term climate changes in isolated habitats.[44] Levels of diversity can be used to indicate the nature of past environments, high levels of diversity and endemism often being thought to have been facilitated by relative environmental stability,[45] the corollary being that intervening areas of relatively low species diversity and endemism have been impacted much more severely by past environmental change. However, one of the problems with this kind of evidence, assuming present-day distribution patterns do carry an imprint of past conditions, is determining when in the past environmental change actually took place. A second problem concerns the assumption that environmental stability in isolated habitats leads to high diversity and endemism, as some biologists are convinced that the opposite is the case.[46]

Numerous biologists working in Africa support the concept that forest was restricted within refuges at the LGM.[47,48,49,50] Refuge theory predicts that forest species of restricted distribution from a wide range of taxonomic affinities should occur together in places where forest survived Pleistocene cooler and drier climates. Frog, snake, mammal, tree, *Begoniaceae* and *Impatiens* distribution records held by the Centre for Tropical Biodiversity (Denmark) show congruent concentrations of high diversity and endemism centred on Mount Cameroon, the Albertine Rift Lakes and the East African mountains.[51] A study of *Begoniaceae* distribution identified three refuge areas in Upper Guinea and a further four smaller forest refuges within Lower Guinea.[52] Tropical moist forest refuges existed in Gabon and the Mayombe region in the Peoples Democratic Republic of Congo (PDRC).

At the LGM the majority of the areas presently supporting tropical moist forest supported dry forest.[53] Bengo and Maley[54] point to differences between Zambezian and Sudanean dry forest; these indicate past isolation across the equator by a band of moist forest, possibly located along the Zaire river system.[55] This is further supported by evidence for a "migratory trackway" between East and West Africa along the Zambezi–Zaire watershed[56] and by the occurrence of distinct sub-species of primates in Central Africa,[57] which were isolated within a "major fluvial refuge".[58] Additional support for Gabon, Cameroon and Central African moist forest refuges comes from the distribution of birds,[59] forest mammals[60] and ethnographic evidence from pygmy populations.[61]

A study of passerine birds showed that centres of species diversity, endemism and disjunction coincide spatially in Ethiopian montane forest, Cameroon/Gabon, east PDRC, and the eastern Tanzanian mountains, the latter extending to the coast.[62] Relatively high diversity within the Albertine Refuge is indicated by a study of forest mammal distribution,[63] flightless insects[64] and molluscs from Kakamega Forest.[65] Similar post-LGM migratory routes out of core areas have been identified for forest tree species throughout Uganda and into neighbouring Tanzania and Kenya.[66] Indeed, within East Africa, many restricted-range tree and shrub species show distinct concentrations.[67] An assessment of ecoclimatic stability based on species distribution indicates that the most stable areas are in the upper Zaire River catchment and on the east-facing escarpments of the East African mountains.[68] Farther to the east, tropical moist forest persisted in parts of coastal East Africa throughout glacial periods due to the moist climate resulting from a relatively constant temperature of the Indian Ocean.[69] A similar importance for forest persistence, attributed to local topography, is indicated for South Africa.[70]

4.4 African Late Glacial Climates

Africa was not strongly influenced by glacial activity at the LGM, with only the high altitudes associated with the High Atlas and Rift Valleys supporting valley glaciers.[71] Glaciers on the Rwenzori mountains reached their maximum extent at 15 000 yr BP although the timing of maximal glacial extent was hetero-geneous on different highland areas.[72] The situation was quite different in Europe, where a single, large, southerly extension of the Scandinavian ice sheet reached approximately 52°N latitude.[73] The ice sheet reached a thickness, at its deepest extent, of some 2500 m.[74] The southern extent of this ice sheet was relatively homogeneous as a result of lack of highland areas about the southern extent to allow for farther extension of the ice. Farther south from this major ice sheet, areas in excess of 2000 m (Alps and Pyrenees) supported ice caps,[75] measuring 500 by 300 km and 300 by 100 km, respectively.[76] Outside of these two main areas of ice cover, a series of valley glaciers were associated with the highland areas of the Balkans, Corsica, Italy and Spain.[77] Although the region was largely ice-free south of the Scandinavian ice sheet, much of the ground was frozen.[78] Thus, the only areas that remained viable for the survival of temperate flora were the three southern peninsulas: Iberian, Italian and Balkan.

Palaeoclimatic estimates for Africa indicate a decrease of 4±2°C relative to the present day.[79,80,81] A wider range of temperature decrease of between −3 and −8°C is suggested for western equatorial Africa.[82] Temperatures in the Nile Delta are estimated to have been between 6 and 7°C below modern levels,[83] whereas winter temperatures in the Saharan Mountains were between 10 and 14°C colder than today.[84] In contrast to these changes, the climate along the Tanzanian/Kenyan coast may have been permanently warm throughout the LGM.[85] Northern African temperatures were about 5°C cooler than the present day.[86] In Central Africa, reductions in precipitation are thought to have been approximately 40% relative to present-day levels.[87] Precipitation levels in coastal Tanzania are thought to have been little changed at the LGM compared to the present day.[88] Indeed, some areas may have been wetter than present at the LGM.[89] A further indication of LGM aridity is indicated by lake levels records at, or about, the LGM; in general these were much lower than the present day.[90] This LGM aridity resulted in the southern extent of the Sahara lying some 5° farther south than present.[91]

4.5   Changing Climate Changing Forests

As a result of the relatively long history of fossil pollen studies in Africa there are more data available on forest history in Africa than on other tropical areas. Changes in the composition and distribution of vegetation inferred from pollen analysis have been well documented in reviews for West Africa,[92] East Africa[93,94,95,96] and for central and southern Africa.[97]

Direct palaeoecological support for the presence of forest refuges comes from West Cameroon; pollen from Lake Baramobi Mbo indicates that the level of tropical moist forest was only slightly reduced at the LGM, whereas pollen from Lake Bosumtwi shows the disappearance of tropical moist forest at the LGM.[98,99] Outside these densely forested areas, tropical moist forest may have persisted as gallery forests along rivers and within valleys.[100] However, Runge,[101] working in an area suggested for the location of a forest refuge

(Kivu province, PDRC), indicates that the area supported open tropical moist forest at the LGM. Pollen from off-shore West Africa indicates a dramatic retreat of tropical moist forest at the LGM, with an associated expansion of dry forest types.[102] Pollen evidence from the Congo delta sediments similarly indicates that tropical dry forest was much more extensive along the rim of the Congo valley at the LGM.[103] Pollen off the Niger Delta is thought to record a savannah corridor between the western (Guinean) and eastern (Congolian) tropical moist forest at the LGM.[104] Thus, tropical dry forest, which presently characterises the Dahomey Gap, was much more extensive at the LGM.

Following an analysis of pollen from six sites along the Western Rift of central Africa, it was found that tropical moist forest was not present at the LGM as discrete forest patches.[105] However, an interesting feature from all the sedimentary records that cover this period is the continued occurrence of tropical moist forest taxa, albeit at reduced levels. This suggests four possible scenarios: 1. tropical moist forest taxa were either present near to, but not within, the catchments so far studied; 2. tropical moist forest taxa were present at relatively low densities within all the catchments; 3. tropical moist forest taxa were present in discrete core areas that have yet to be delimited; or 4. pollen was transported long distances into sedimentary basins from tropical moist forest at lower altitudes. Sites at lower altitude do not support the last suggestion: studies from Lake Mobutu Sese Seko[106] and Lake Tanganyika[107] indicate open grassland with isolated forest patches at the LGM. Within the relatively low-lying areas of the eastern Rift Valley of Kenya the vegetation was dominated by tropical dry forest, although elements of forest taxa were present close to the lake margins.[108]

Areas where forests are believed to have persisted at the LGM under maximum climate change are now known as forest refuges. The theory of forest refuges was developed largely from results of investigations in South America,[109] which have since been added to, following further studies in the neotropics and in other parts of the world.[110] Although there is evidence in support of the existence of forest refuges in some areas,[111] the evidence is largely circumstantial and based mainly on present-day distributions of plants and animals, the output of coarsely resolved biome response models or isolated fossil-based studies. Indeed, direct evidence in support of the theory, in the form of well-dated fossil remains *in situ*, remains lacking for most parts of the world and hence the controversy continues. Thus, although there is general acceptance that what are now forest taxa must have survived the LGM somewhere, exactly where this survival took place and the nature of vegetation within those refuge areas remain subjects for debate.

4.6 Past Climate Change as a Predictor of Diversity

The proposed climate-induced reduction of area of the Africa forest during the Pleistocene is a possible explanation for the relative poverty of African biodiversity compared to the other two main areas of tropical forest in South

America and Indo-Malaysia. This refugium hypothesis has also been applied to South America, though rather more controversially than in Africa. Patterns of species richness in South American forests have been attributed to Pleistocene refugium and the isolating mechanism of periodic forest withdrawal into refugia has been used as an explanation of species richness through an "evolutionary pump" of isolation and coalescence.[112,113,114] This is in contrast to Africa, where the same process of forest reduction has been used to explain species poverty rather than richness. The tide of scientific opinion has now swung away from Pleistocene climate-change being a key determinate of South American forest diversity,[115,116] with the evidence pointing more towards a straightforward ecological, rather than historical, explanation with the highest diversities correlated with high rainfall, short dry season and younger fertile sediments.[117] In Indo-Malaysia the evidence also suggests that Quaternary climate changes have had little impact on lowland tropical forests as the region is buffered by the close proximity of the ocean almost everywhere, though there is high inter-annual variability in rainfall due to climatic fluctuations caused by sea surface temperature changes of the El Niño Southern Oscillation.[118]

The Pleistocene refugium hypothesis provided the first scientific basis for localising areas of high species diversity, the "hotspots" in tropical forests. If these areas could be located and special attention paid to their conservation, then threats to tropical forest biodiversity could be minimised as they represented the places where forests had survived periods of past reduction in forest extent. For example, in South America a series of reserves were planned to coincide with proposed refugia.[119] However, as the controversy outlined above over location of the Amazonian refugia indicates, the refugium hypothesis as an explanation for observed pattern of species is not necessarily straightforward. An alternative explanation proposed for eastern Africa is that some areas are geologically and climatically stable over evolutionary time periods, thus allowing species to survive and differentiate into the distinct morphological types that we recognise as species.[120,121,122] This stability hypothesis therefore suggests that the high species diversity and endemism in these hotspots is not due to extinction outside the hotspot, but from accumulation of species within it. This has important management implications, because species in these centres of ecological stability will be adapted to lack of disturbance. Management interventions which cause disturbance will then lead to a loss of species adapted to stability and replacement with more widespread species which can cope with a range of ecological conditions. This is in contrast with the dynamic nature of species associated with the refugium hypothesis, as these species will be restricted to refugia and then disperse readily back into suitable habitats when the weather becomes wet and warm again. The stability hypothesis has recently gained empirical support through analysis of pollen from cores taken from a swamp in the Udzungwa mountains of Tanzania, which are part of the Eastern Arc tropical forest biodiversity hotspot. Remarkably, the core shows relatively little change in forest composition during the last glacial maximum in contrast to similar cores taken elsewhere in eastern Africa.[123] Thus the nature of threats to tropical forests will vary according to past history. Forests that have a long history of change will be more resilient to disturbance than those that have evolved under conditions of comparative ecological stability. In addition, understanding the responses of tropical forests to climate change in the past will help us to understand the potential impacts of future climate change which is regarded as one of the major future threats to biodiversity.[124,125]

5 Reasons for Local-scale Variation due to Present-day Ecology

The shifting of continents and global changes in climate associated with the ice ages are responsible for the basic patterns of tropical forest diversity that we see today. However, in addition to these large-scale processes, there is a great deal of local-scale variation. Rainfall is a key determinate of the level of diversity both in overall annual levels of precipitation and seasonality. Biodiversity is greatest in forests with high rainfall and no marked dry season.[126] This relationship helps to explain the difference in levels of diversity between the main continental areas of tropical forests. Much of tropical South America and the Indo-Malayan archipelago have a per-humid climate with greater than 100 mm of rain in every month of the year. In contrast, few places in Africa have a per-humid climate, with even high rainfall areas experiencing marked dry seasons. For example, the peak of Mt Cameroon is one of the wettest places in the world with an annual rainfall of about 10 000 mm, but there is still a dry season in December and January. Exceptions to this are found in the biodiversity hotspots; for example, the Usambara and Uluguru Mountains in the Eastern Arc hotspot.

Mountains are also associated with high diversity. There are a number of reasons for this. First, mountains are often associated with high rainfall caused by warm moist air cooling as it rises. Second, temperature and moisture gradients on mountains create a wide range of different habitats and on most wet tropical mountains there is almost a complete turnover of species from low to high elevations. Third, under conditions of climate change, plants and animals can migrate along the environmental gradients and so avoid local extinction. Fourth, clearance of forests for agriculture is likely to be greatest in flat, easily accessible areas, resulting in high-diversity forest on the steep-sloped mountains, among a sea of agriculture (see below). If continuity of forest cover over mountains is disrupted by human activities such as agriculture, then the environmental gradients are disrupted and the potential is lost for mountains to act as buffers to climate change.

Other reasons for local-scale maintenance of diversity include heterogeneity in soils and groundwater; pest pressure under which seedlings fail to regenerate near their parents because of the pest load carried by adults; and intermediate-level disturbance that is large enough to create new habitats for species to enter a community, but not so great as to cause major changes. Remarkably, although tropical forests are rightly famed for their high diversity, some forests are characterised by mono-dominant stands of a single species such as

Gilbertiodendron dewevrei in the central African Ituri forests. There are several possible explanations for this phenomenon. It could be due to seasonal flooding or be part of a successional stage following major disturbance. Mono-dominance in tropical forests might also be a function of the species itself, with the adults casting deep shade and forming a deep leaf litter layer and so preventing seedling regeneration. Alternatively, the species may have poorly dispersed seeds or be a mast fruiter, producing huge quantities of seed and so causing pulses of seedling establishment.[127] But perhaps the most interesting possibility is that the species forming mono-dominant stands also tend to be those with ectomycorrhiza. This is a fungal association with the tree roots, helping the plant gain soil nutrients by forming a "Hartig" net around the stunted root tips and penetrating into the root cortex. The ectomycorrhizal habit is found in particular taxonomic groups of plants such as the *Caesalpiniaceae* and *Dipterocarpaceae* (Pinus, Fagaceae in temperate regions). If the main cause of mono-dominance is a major disturbance then this has important implications for management of tropical forests as the effect of the disturbance is a long-lasting reduction of diversity.[128] In the case of the Ituri *Gilbertiodendron* forests charcoal over 2000 years old was recovered from pits dug in the forest floor suggesting a different tree species composition and that fire might have been the trigger to initiate mono-dominance.

6   Past Anthropogenic Impact on Tropical Forests

There is a long history of human impact on tropical forests, particularly through the use of fire to transform closed forest formations into grasslands and woodlands that are more suitable for large mammals and domestic stock, and more recently for clearance for agriculture. Some human societies live inside tropical forests, perhaps most famously the central African forests peoples known as "pygmies". However, wild food resources are limited, so population densities are low and people are usually associated with rivers and clearings rather than the deep shaded forest. Their impact on the natural forest ecosystem is therefore low. In marked contrast, people who live outside the forest have historically used fire to literally "terraform" the landscape to make it more economically productive.[129]

The modern extent of closed forest in Africa is largely determined by fire and large mammal browsers.[130,131] Fire use in Africa has a very long history, though much of the evidence for early fire use is inferential rather than direct.[132] The oldest suggested use of fire was 1.0–1.5 million years ago, based on deposits from the Swartzkrand cave in South Africa.[133] Marine sediments on the Sierra Leone rise off the west African coast show that fire incidence was relatively low until about 400 000 years ago when vegetation fires increased, particularly during the periods when global climate was changing from inter-glacial to glacial.[134] Outside of Africa, there is evidence for controlled use of fire by humans in Israel 790 000 years ago[135] and association between human activity and fire in China 500 000–200 000 years ago.[136] In Indo-Malaysia, there is presence of charcoal in marine sediments from north of New Guinea dating from 52 000 years ago and vegetation changes in Sulawesi around 37 000 years ago are considered to be due to burning rather than climate change *per se*.[137] At Lynch's Crater in tropical north-eastern Queensland sediments indicate burning starting around 45 000 years ago and are not correlated with climate shifts and there is no evidence for sustained changes before this time in a record that goes back 220 000 years.[138] This suggests human-induced burning was responsible for a major change in vegetation in the area from rainforest to sclerophyll woodland. Elsewhere in tropical Australia, burning increased in the Kimberleys 130 000 years ago with major changes about 46 000 years ago.[139] In the high-altitude forests of the South American Andes near Lake Titicaca there is evidence for human disturbance of vegetation dating from about 3100 years ago[140] and it is thought that the sharp demarcation between Andean forest and grassland is due to millennia of human-induced burning.[141]

In Africa, the first indications of settled agriculture are from about 8000 years ago. These include settlements near the Nile and linguistic evidence, including agricultural terms elsewhere in tropical Africa.[142] Agriculture spread into tropical Indo-Malaysia around 5000 years ago, with evidence for rice-growing in Sumatra and taro root crops in New Guinea uplands.[143] South American agriculture is at least 7000 years old, emerging in the highlands and spreading to the lowlands.[144] Intensification of agriculture about 3–4000 years ago is associated with deforestation on all tropical continents. However, a major expansion of extent of the oil palm *Elaeis guineensis* in west Africa around 2000 years ago, formerly thought to be due to agriculture, is now considered to be the result of climate change.[145] Today, burning and replacement of forest by agriculture are major threats to tropical forests.

7 Present Anthropogenic Impact and Management of Tropical Forests

Although, historically, humans have had a major impact on forests, technical innovations in logging and mechanised capital-intensive methods, requiring fast returns on investment, have meant that forest conversion and degradation have increased in recent decades. The rise in threats to tropical forests and increasing public concern over the effects on biodiversity during the 1980s was a contributing factor to formulation of the Convention on Biological Diversity at its launch at the Rio de Janeiro Earth Summit in 1992. However, this did not halt logging, which not only degraded forests but also made them more susceptible to fire.[146] Controversially, some studies reported that, although logging reduced the density of trees, the number of stems of trees of different species did not decline following logging,[147] suggesting that logging might not have the devastating effect predicted by conservationists if post-logging management can help the forests recover. The difficulty here is the ability of forest managers to apply suitable post-logging treatments. Whilst the science of forest restoration is well established for temperate and boreal regions,[148] management of tropical forests following logging has been problematic with few, if any, success stories.[149]

The main reason for logging tropical forests is commercial gain. It has been argued that timber companies are granted concessions to exploit forests at a price below the cost of subsequent effective post-logging management.[150] The potentially renewable forest resource is thus "mined" for its old-growth values rather than managed for its ability to regenerate. Distortions in the economics of tropical land-use also lead to deforestation and replacement of species-rich forests with agriculture.[151] Some efforts to correct the market failures that led policy makers to undervalue tropical forests included estimation of the monetary value of non-timber forest products (NTFPs), such as edible fruits, oils, latex, fibre and medicines. The values of NTFPs were compared with those of major forest products, including saw-logs and pulp-wood. One of the early studies showed that in one hectare of species-rich Amazonian forest, the total net revenues generated by sustainable exploitation of minor forest products were substantially higher than those resulting from forest conversion.[152] The NTFPs generated a net present value (*i.e.* discounted future returns) of \$6330 ha^{-1} compared to \$490 for timber.

Although this and subsequent studies suggest that one of the major threats to tropical forest is the failure of policy makers to adequately take into account the real values of NTFPs, there are some problems with this type of economic approach. First, the high values of minor forest products assume a strong social and economic linkage between people living near the forest and the forest's ecology. In fact, what tends to happen is that as societies advance economically they rely less on multiple NTFPs, preferring instead to obtain household goods, medicines and food from external sources. This means that NTFPs are "sustainable as long as underdevelopment, economic stagnation, unemployment and low wages persist".[153] Second, NTFPs with high values and commercial potential tend to be "captured to culture" and introduced into agriculture. When this happens, wild sources of the crop lose their value, as the costs of gathering from native forests are higher than harvesting from cultivation. Third, not all tropical forests have high NTFP values.[154] Many forests are not used extensively for extraction of NTFPs and the only way to increase their values to justify prevention of replacement by agriculture is by including existence values. Existence values are the values that people put on the simple existence of something such as a "grand scenic wonder",[155] but are highly controversial as it is not clear whether they can actually be converted into monetary values.

The hydrological functions of tropical forests are also regarded as being of high value and many forest reserves, particularly on mountains, were initially established as "catchment forest reserves" to preserve water supplies. Forests provide hydrological environmental services through regulation of droughts and floods, control of soil erosion and amelioration of climate and ground water recharge.[156] The multi-layered vegetation structure prevents direct impact of heavy tropical rains on the soil, stopping soil erosion caused by splashing and slowing surface runoff. Instead, the rain is intercepted by the

Table 3   Rainfall and evapotranspiration (ET) in millimetres a year from a range of tropical forests. Evapotranspiration is from: evaporation of precipitation intercepted by the vegetation; transpiration and evaporation from the ground layer.

Location	Elevation	Rainfall mm yr^{-1}	ET mm yr^{-1}
Colombia	1150	1985	1265
Costa Rica	2400	2695	365
Indonesia	1750	3305	1170
Malaysia	870	2500	695
Philippines	2350	3380	390
Venezuela	2300	1575	980

canopy and tends to flow down stems or drip on to the forest floor, which is covered by protective leaf litter. Roots bind the soil, preventing erosion and assisting infiltration of water to sustain ground water supplies. Forests ameliorate local climates by covering and shading soils so that the forest understorey maintains relatively even temperatures and a high humidity. In addition, a high percentage of rain returns to the atmosphere through direct evaporation from canopy surfaces and via transpiration of groundwater up through the trees to the leaves (Table 3).

After a rain storm the forest canopy is shrouded in mist and cloud as water evaporates, helping to retain a locally humid climate. Condensation of cloud on vegetation surfaces can be an important source of precipitation, supplementing that arriving through rain. This is particularly true in montane forests with heavy epiphyte loads as epiphytic plants growing on the trees in the canopy substantially increase the surface area available for condensation. Called "horizontal" or "occult" precipitation, the volume of this source of water is difficult to estimate. Annual totals of horizontal precipitation estimated with fog-catchers range from 70 mm at 3100 m elevation in Venezuela to 940 mm at 1300 m elevation in eastern Mexico.[157]

Hydrological services provided by forests are adversely affected by logging and forest clearance, most obviously by removal of vegetation and alteration of the structural characteristics of the forest. This then affects the impact of rain on the soil, infiltration, humidity and horizontal precipitation. Timber extraction also leads to soil compaction on log landings and skidding trails, which results in a decline in soil pore space and infiltration rates, and an increase in runoff and likelihood of land slips on steep slopes.[158] Another of the threats to tropical forests is the difficulty of linking the hydrological values of forested catchments with the downstream benefits. Loss of forest or extensive logging leads to higher runoff rates, changes in flooding patterns and therefore loss in agricultural production. The problem is that catchment protection leads to economic losses to hill-farmers and forest owners, whereas hill-farming and logging lead to economic losses to downstream paddy-farmers.[159] As yet there have not been any effective ways of dealing with this equity issue and it remains one of the major challenges of tropical forest management.

8   Case Study: Management of the Mufindi Forests

A large proportion of biodiversity is maintained in economically productive landscapes. Biodiversity conservation thus needs to be compatible with land use that leads to positive financial gains. Developing countries are rich in biodiversity, but are not wealthy enough to provide conservation compensation payments such as those used in developed countries. Therefore, an important research area is to find practical ways of implementing sustainable and equitable biodiversity conservation at low cost to the businesses that support it. The need for business to engage in biodiversity conservation is recognised by the Convention on Biological Diversity (CBD) through its Business and the 2010 Biodiversity Challenge meetings. These recommend that companies need to define and implement clear strategies for biodiversity conservation. Sector-specific good practise guidelines aligned with the CBD are seen as an important way forward, including guidance on how industry should co-operate with local communities. Maintaining biodiversity in economically productive landscapes is also a highly effective way of meeting the Millennium Development Goal of environmental sustainability.

Despite the many threats and unsolved problems in tropical forest management, there are some success stories. Here we describe management of the tropical montane forests on the Unilever tea estate at Mufindi in the Udzungwa Mountains of Tanzania, which are part of the Eastern Arc range. The Eastern Arc is an ancient crystalline chain of mountains of the Mozambique belt under the tropical Indian Ocean climatic system. The forests range in elevation from sea level to 2400 m with a seasonal to perhumid climate and rainfall up to 4000 mm year^{-1}. They are highly fragmented due to topography and disturbance (Figure 1). Together with the Coastal forests on sedimentary rocks of the coastal plain, they form a centre of biological endemism recognised as one of the top 25 biodiversity hotspots.[160]

The plant species endemism is around 30% of the flora, with endemic species being biogeographic relicts, phylogenetic relicts and neo-endemics. The presence of relictual species suggests that the Eastern Arc forests have been in existence for tens of millions of years under a long-term stable geology and climate.[161] A stable ecosystem over an evolutionary time-period would result in Eastern Arc plants being adapted to lack of disturbance, a suggestion that is given some credence by the loss of restricted range species following disturbance (Table 4).

The 1998 Tanzania Forest Policy recognises the importance of biodiversity conservation. Article 18 of the policy states that "Biodiversity conservation and management will be included in the management plans for all protection forests. Involvement of local communities and other stakeholders in conservation and management will be encouraged through joint management agreements." Legal protection to individual species, such as those on the IUCN "Red List", is afforded by the 2002 Forest Ordinance, which is the legal instrument supporting the policy. Currently about 191 Eastern Arc plant taxa are red-listed, and a further 986 endemic plant taxa are potentially threatened.

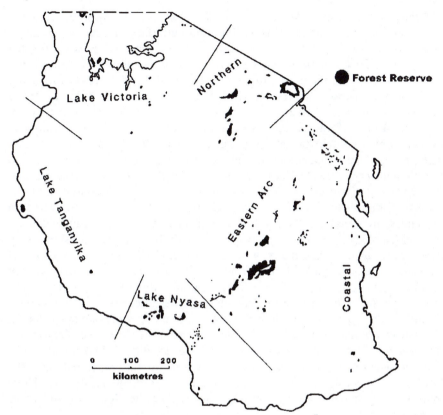

Figure 1 Extent of the forests in Tanzania as indicated by presence of forest reserves with the areas of forest divided by geology and climatic influence.[162]

Table 4 Numbers of tree species in samples of 60 trees of $\geq$ 20 cm diameter at breast height from a range of montane forests in the Eastern Arc. The southern Udzungwa forests are structurally similar to the other forests, but are growing on a site of ancient cultivation and have much lower species diversity and no endemics.[163]

Location	Species	Endemics
West Usambara	26	5
Southern Nguru	21	4
Northern Udzungwa	20	6
Southern Udzungwa	9	0

To develop compliance with the new forest policy and law, Unilever Tea Tanzania Limited (UTTL) have developed a management strategy for their long-term leasehold of nearly 20 000 ha of land, owned by the Government of Tanzania, in the Mufindi area of the Udzungwa Mountains, to the southern end of the Eastern Arc.

Many of the Eastern Arc endemic plants occur on the Mufindi estate. In addition, rare birds, amphibians, reptiles and butterflies with restricted ranges are known also to occur. In Mufindi, 15% of the land is used for tea production, 20% represents land converted to other uses, including timber for construction on the estate, fuel wood, residential and other infrastructure and facilities. Approximately 65% of the estate is covered by relatively undisturbed forests, wetlands and grasslands. Six main habitats can be identified in Mufindi, from east to west: 1. Escarpment forests, along the Luisenga Stream which contain the globally threatened flycatcher, the Iringa Akalat (*Sheppardia lowei*); and very rare shrubs, including members of the bamboo, myrtle and witchhazel families – respectively *Hickelia africana*, *Eugenia mufindiensis* and *Trichocladus goeztei*. In addition, a rich terrestrial and epiphytic orchid flora is found here. At least 67 tree and shrub species, including endemics, are utilised. 2. Plateau forests; endemic animals and plants including the Iringa Akalat. 3. Plateau grasslands, with large populations of many terrestrial orchid species, many of which, including *Disa* sp. *Satyrium* sp. and *Habenaria* sp., are under threat from high-volume trade for food between Tanzania and Zambia. 4. Plateau forest patches; these are important "stepping-stone habitats", providing corridors between forests, and are habitat islands in the grasslands providing a food source and nesting habitat for birds. 5. Plateau woodlands, where the small spiny succulent *Euphorbia caloderma* is found and which is not known from any other locality. 6. Plateau wetlands, with a rich orchid flora. Also of importance are the converted habitats, including tea, eucalyptus, black-wattle (*Acacia mearnsii*) and road edges. These may provide valuable feeding, passage and shelter sites for birds and other species.

Human pressure is mounting on the Mufindi forest. Approximately 15 villages within 6 legislative wards of Mufindi with a total human population of 100 000 are located within 10 km of the estate's boundaries. Thirteen primary and two secondary schools with a total of 8000 students are also located within 10 km of the Mufindi estate. In addition, 7000 people are employed by UTTL. They live both within (the majority) and outside the estate's boundaries. While the practical evidence is that plant resources, particularly trees and shrubs, are collected from the forest in increasing quantities for a wide variety of uses, it is not well known what impact this is having on target and non-target species. The extent to which households are dependent on these natural resources either for cash income or for subsistence is also unclear.

Human impacts are still at a comparatively small scale, but are increasing. Therefore, an ideal opportunity is presented to develop an understanding of the value of the Mufindi estate's resources amongst user and other communities, before pressure on the estate's resources becomes unsustainable. Since 2000, UTTL has been developing and implementing a Biodiversity Action Plan. This is in line with Unilever's global requirements that ultimately all producers of tea, palm oil, spinach, peas, tomatoes and edible oils must apply ten sustainability indicators. These include protection of biological diversity, support for the local economy, capacity building for suppliers of raw materials and sharing of knowledge and good practice.

Nearly one decade ago, Unilever developed sustainability guidelines for the sourcing, on a global basis, of many of the raw materials it uses in its food and home and personal care products. These are now being increasingly applied by producers, for example, of vining peas (UK); spinach (Italy, Austria, Germany); tomatoes (Brazil, Australia, Greece, California); palm oil (Malaysia, Ghana, Indonesia); tea (Kenya, Tanzania, India, Sri Lanka); olive oil (Greece and other Mediterranean suppliers); and in the future sunflower and rapeseed oil. The ten sustainability indicators being implemented by producers are (with specific reference to tea): 1. Soil fertility (addressing organic matter, soil compaction, soil pH and salinity); 2. Soil loss (addressing soil erosion, ground cover and top soil use for nursery); 3. Nutrients (ratio of exports to inputs; nitrogen input from biological fixation, loss of nitrate and phosphate by surface runoff, sediment erosion and to ground water); 4. Pest management (arthropod pests and fungal diseases, pesticide use, weed control); 5. Biodiversity (crop genetic diversity, biodiversity without and around the estate); 6. Product value (profitability, product quality); 7. Energy (efficiency and the use of renewable resources, reducing GHG emissions); 8. Water (irrigation, factory process water, water harvesting and the sustainability of water supply); 9. Social and human capital (relationships, human capital); 10. Local economy (use of local management and worker capacity; use of local suppliers *etc.*). Unilever has therefore completed nearly a decade of work on sustainability indicators for the production of an increasingly wide range of crops which has influenced the practice of thousands of suppliers. In Kenya, the company is now working with over 300 000 small-holder suppliers of tea on long-term use of sustainability practices.

UTTL has introduced its staff, employed in the field and in other sections of the company, as well as its small-holder suppliers, to the importance of the estate for unique animal and plant life and ecological services, including water supply and soil structure, and has trained some staff in research techniques, particularly in the application of the species and habitat monitoring protocols. A fundamental aspect of this work is to identify more clearly the following, and in the light of the answers to the analysis, develop and begin to implement a practical programme to alleviate user pressure on the most sensitive species in the Mufindi estate:

- Which human communities have most impact on Mufindi's six habitats and individual herbaceous or woody plants?
- What plants are most used; how widespread are they and what are their populations; are the most used species also the most threatened?
- What are they used for? Are there opportunities to (a) reduce pressure on wild populations by domestic propagation of the same or similar species and (b) introduce more sustainable ways of using the forest resource?
- Is there scope for planting schemes to provide corridors and feeding areas for species?

The objectives of the biodiversity management plan are:

- To secure protection long-term for animal and plant "species of concern" in the Mufindi forest "hot-spot", by:
    - continuing research on "species of concern" on the Mufindi Estate; their status, range, *etc.*
    - providing alternative sources of forest products for human use, by agro-forestry and other activities involving plantings of desired shrubs, trees and herbaceous plants
    - determining whether sustainable use of naturally growing forest resources is feasible and whether community-developed management plans are a practical option; and to begin implementing these if the conclusion is positive
    - encouraging an understanding by all stakeholders of their dependence on forest resources and ecosystem services, and therefore their co-operation in long-term sustainable management
- To apply Unilever's ten sustainability indicators for agriculture (which include protection of biodiversity) to the production of tea by small-holders and others, and achieve integration of these practices with small-holder commitment to undertake resource-use practices which relieve pressure on the forest, either by planting required species or by sustainable use of the natural forest.
- To use the results of this project to strengthen Unilever's sustainability practices in relation to ecosystem management in other areas of the company's operations.
- To use the results of this project to influence other companies in terms of their sustainability practices, in the food and other sectors; for example, through the Sustainable Agriculture Initiative, begun by Unilever and now involving *c.* 20 global companies.

The expected results of implementation of the management plan are:

1. Reduction of impacts on key forest areas and identified "species of concern" therein, with neither short- nor long-term negative cultural or economic consequences for the livelihoods of local people.
2. Local communities committed to contributing to the company's efforts to reduce pressure on the biodiversity "hot-spot" areas of the estate and to managing their use of natural resources in a sustainable manner, thereby contributing long-term to their own economic welfare.

The management plan deliverables will be:

- Reduced impact on the most important "hot-spot" species, in collaboration with user-communities, by developing alternative sources of needed products
- Improving status and long-term prospects for survival of "hot spot" species on the Mufindi estate.

The Mufindi management plan has been developed with the intention of offering a model for future application by other tropical conservation management projects. If successful, the Mufindi case study will illustrate that protection of forest biodiversity is compatible with economic production and that, with adequate planning and commitment, a commercial firm operating under conservation-minded government policy and legislation can work with local communities to reduce threats to biodiversity.

9 The Future

Threats to tropical forest biodiversity are many-fold: fire, causing a shift to grassland, conversion to agriculture, logging and forest policies that do not recognise the value of non-timber forest products. However, it is now recognised that climate change induced by human activity is the major future threat to tropical forest biodiversity and will be compounded by the other threats described in this chapter. A recent model of the change in patterns of African plant diversity showed a massive loss of suitable climate for forest species in the area currently occupied by tropical forest in west and central Africa.[164] If the model is correct, then global warming will bring massive deforestation to this region. But should we be worried? We have already described the massive changes that occurred to tropical forests following climate shifts in the last glacial maximum. For example, the changes predicted for the African forests may have happened in the past, caused by a major southern shift in the dry Sahara climate.

Future climatic change is likely to be different from the past events discussed earlier for two reasons. First, the rate of predicted climatic change exceeds that of past climatic change and, second, many natural habitats have become fragmented by human populations, producing isolated habitat islands that are unable to migrate. There is little dispute that global climates are changing, and the nature of this change is projected to continue even if the most extreme abatement scenarios are implemented.[165] The process of forest conservation, through the formation of protected areas, is based on the principle of preserving habitats for future generations. For this principle to be successful it is necessary to investigate a plethora of issues surrounding areas that are now foci for protected area status. Although the primary agent of change over the past few hundred years has been direct human activity, there is little doubt that future climate change will impact on forest composition and distribution. Although the specifics of past climate change are very different from those suggested for the future,[166] the threat to biodiversity posed by global climate change is recognised.

Future climate and environmental change is predicted to cause major changes to biodiversity, for which new conservation paradigms must be established that need predictions of potential future change on which to base conservation strategy.[167] A switch from static protected-area management to dynamic management systems that account for climate-induced migration of species is needed, but this must take into account the ecosystem history and causes of species richness. Knowledge of potential impacts will enable policy-makers to prepare appropriate strategies in advance of climate-change events, and so minimise and manage adverse effects. Only through a more complete understanding of impacts and interactions of climate change on ecosystem functioning can the likelihood of potential future scenarios be estimated and so appropriate policy prepared. This can be driven by understanding our need for services that ecosystems provide.

With increasing recognition of the impacts of climate change on ecological, social and economic levels, there is a need to develop a science-led policy framework. To develop such a laudable aim, there is a need to develop research capacity for communities to contribute to and benefit from this process. Indeed, the lack of research capacity in tropical forest countries can be counted as a major threat to tropical forest biodiversity because indigenous expertise is needed to guide local policy-makers. For example, European researchers have had a long-term focus on African ecosystems and there are a number of well-developed and respected research groups with an African focus within most European member states. However, these have largely been a result of "pioneering" research collaborations and often lack integration both at methodological and spatial levels.

Climate change and subsequent ecosystem response is of major importance to policy makers, but it is an area surrounded by uncertainty and controversy. To link findings of pure research, policy and economics, the void in understanding natural and historical processes behind present-day landscapes needs to be filled. When there is sufficient information it will be possible to move away from reactionary response and management of many urgent environmental and development issues. Increased scientific understanding regarding land use, soil and water conservation, climate change, capacity building and the wider socio-economic consequences of climate change through likely changes in ecosystem form and function need to be understood for long-term sustainable development. By reconstructing past impacts of climate change in relation to potential future events, it is possible to make an assessment of future risks, thereby helping to guide current policy on the impacts of climate change locally, with manifestations regionally and indeed globally. Given the growing and tangible impacts of climate change, new international relationships will be fostered and developed that will become increasingly important as policies on managing the consequences of global climate change move from the national to the international political arena. At the heart of this is a realisation that numerous complimentary research strands need to be woven together to form a complete understanding of ecosystem dynamics and response to environmental change; only then will findings on the magnitude of ecosystem change and associated societal impacts be able to move from the scientific to the policy arena.

References

1. J. E. M. Baillie, C. Hilton-Taylor and S. N. Stuart, "The IUCN Red List of Threatened Species", IUCN, Gland, 2004.

2. P. Mayaux, P. Holmgren, F. Achard, H. Eva, H. -J. Stibig and A. Branthomme, *Phil. Trans. Biol. Sci.*, 2005, **360**, 373.
3. F. Achard, H. D. Eva, H. -J. Stibig, P. Mayaux, J. Gallego, T. Richards and J. -P. Malingreau, *Science*, 2002, **297**, 999.
4. G. P. Asner, D. E. Knapp, E. N. Broadbent, P. J. C. Oliveira, M. Keller and J. N. Silva, *Science*, 2005, **310**, 480.
5. World Resources Institute, "World Resources 1992–1993", Oxford University Press, New York and Oxford, 1992.
6. T. C. Whitmore and J. A. Sayer, "Tropical Deforestation and Species Extinction", Chapman and Hall, London, 1992.
7. T. C. Whitmore, "An Introduction to Tropical Rain Forests: Second Edition", Oxford University Press, Oxford, 1998.
8. R. S. Duncan and C. A. Chapman, *Restor. Ecol.*, 2003, **11**, 198.
9. N. Myers, R. A. Mittermeier, C. G. Mittermeier, G. A. B. da Fonseca and J. Kent, *Nature*, 2000, **403**, 853.
10. R. Good, "The Geography of Flowering Plants", Longmans Green and Co., New York, 1947.
11. J. C. Morrison, D. M. Olson, C. J. Loucks, E. Dinerstein, T. F. Allnutt, E. D. Wikramanayake, T. H. Ricketts, N. D. Burgess, Y. Kura, G. V. N. Powell, J. F. Lamoreux, E. C. Underwood, W. W. Wettengel, J. A. D'Amico, P. Hedao, I. Itoua, K. R. Kassem and H. E. Strand, *BioScience*, 2001, **51**, 933.
12. R. Govaerts and J. Dransfield, "World checklist of palms", Royal Botanic Gardens Press, Kew, 2005.
13. P. H. Raven and D. I. Axelrod, *Ann. MO Bot. Gard.*, 1974, **61**, 539.
14. W. C. Pitman III, S. Cande, J. Labrecque and J. Pindell in "Biological relationships between Africa and South America", P. Goldblatt (ed.), Yale University Press, New Haven and London, 1993, 15.
15. M. E. Rasanen, J. S. Salo and R. J. Kalliola, *Science*, 1987, **238**, 1398.
16. T. C. Whitmore, "Wallace's Line and Plate Tectonics", Clarendon Press, Oxford, 1981.
17. E. S. Vrba in "Biological Relationships between Africa and South America", P. Goldblatt (ed.), Yale University Press, New Haven and London, 1993, 393.
18. A. D. Pan, B. F. Jacobs, J. Dransfield and W. J. Baker, *Bot. J. Linn. Soc.*, 2006, **151**, 69.
19. J. A. Coetzee in "Biological Relationships between Africa and South America", P. Goldblatt (ed.), Yale University Press, New Haven and London, 1993, 37.
20. B. F. Jacobs and P. S. Herendeen, *Palaeogeography, Palaeoclimatology, Palaeoecology*, 2004, **213**, 115.
21. M. R. Silman in "Tropical Rainforest Responses to Climatic Change", M. B. Bush and J. R. Flenley (eds), Springer, Berlin, 2006, 269.
22. R. A. Marchant and H. Hooghiemstra, *Earth Sci. Rev.*, 2004, **66**, 217.
23. K. J. Willis, *Endeavour*, 1996, **20**, 110.
24. W. R. Peltier, *Science*, 1994, **265**, 195.

25. L. G. Thompson, T. Yao, M. E. Davis, K. A. Henderson, E. Mosley-Thompson, P. -N. Lin, J. Beer, H. -A. Synai, J. Cole-Dai and J. F. Bolzan, *Science*, 1997, **276**, 1821.

26. W. S. Broecker, *Proc. Natl. Acad. Sci. USA*, 2000, **97**, 1339.

27. H. A. Osmaston in "Quaternary and Environmental Research on East African Mountains", W. C. Mahaney (ed.), A. A. Balkema, Rotterdam, 1989, 7.

28. T. F. Stocker and O. Marchal, *Proc. Natl. Acad. Sci. USA*, 2000, **97**, 1362.

29. K. J. Willis, L. Gillson and T. M. Brncic, *Science*, 2004, **304**, 402.

30. J. C. Stager, P. A. Mayewski and D. L. Meeker, *Palaeogeography, Palaeoclimatology, Palaeoecology*, 2002, **183**, 169.

31. R. B. Dunbar, *Nature*, 2003, **421**, 121.

32. R. A. Keer, *Science*, 2003, **299**, 183.

33. T. C. Johnson, E. T. Brown, J. McManus, S. Barry, P. Barker and F. Gasse, *Science*, 2002, **296**, 113.

34. S. L. Lewis, Y. Malhi and O. Phillips, *Phil. Trans. Biol. Sci.*, 2004, **359**, 437.

35. C. D. Thomas, A. Cameron, R. E. Green, M. Bakkenes, L. J. Beaumont, Y. C. Collingham, B. F. N. Erasmus, M. F. De Siqueira, A. Grainger, L. Hannah, L. Hughes, B. Huntley, A. S. Van Jaarsveld, G. F. Midgley, L. Miles, M. A. Ortega-Huerta, A. Townsend Peterson, O. L. Phillips and S. E. Williams, *Nature*, 2004, **427**, 145.

36. I. C. Prentice, *Quaternary Res.*, 1985, **23**, 76.

37. R. A. Marchant and D. M. Taylor, *New Phytologist*, 2000, **146**, 505.

38. S. Hicks, H. Tinsley, A. Huusko, C. Jensen, M. Hättestrand, A. Gerasimedes and E. Kvavadze, *Review of Palaeobotany and Palynology*, 2001, **117**, 186.

39. M. J. Bunting and D. Middleton, *Quaternary Res.*, in press.

40. S. T. Jackson, R. S. Webb, K. H. Anderson, J. T. Overpeck, T. Webb III, J. W. Williams and B. C. S. Hansen, *Quaternary Sci. Rev.*, 2000, **19**, 489.

41. D. Jolly, D. M. Taylor, R. A. Marchant, A. C. Hamilton, R. Bonnefille, G. Buchet and G. Riolett, *J. Biogeogr.*, 1997, **24**, 495.

42. D. H. Urrego, M. R. Silman and M. B. Bush, *J. Quaternary Sci.*, 2005, **20**, 693.

43. Brown and K. S. Jnr, *Publ. Lab. Zool. École Norm. Sup.*, 1976, **9**, 118.

44. J. Fjeldså and J. C. Lovett, *Biodiv and Cons.*, 1997, **6**, 315.

45. J. C. Lovett and I. Friis in "The Biodiversity of African Plants", L. J. G. van der Maesen, X. M. van der Burgt and J. M. van Medenbach de Rooy (eds), Kluwer Academic Publishers, Dordrecht, 1996, 582.

46. M. B. Bush, *Biol. Conservat.*, 1996, **76**, 219.

47. J. Kingdon, "Island Africa", Collins, London, 1990.

48. M. Colyn, A. Gautier-Hion and W. Verheyen, *J. Biogeogr.*, 1991, **18**, 403.

49. M. Rietkerk, P. Hetner and J. J. F. E. de Wilde, *Bulletin Museum Natural History Paris Section B Adansonia*, 1995, **17**, 95.

50. M. Sosef in "The Biodiversity of African Plants", L. J. G. van der Maesen, X. M. van der Burgt and J. M. van Medenbach de Rooy (eds), Kluwer Academic Publishers, Dordrecht, 1996, 602.

51. J. C. Lovett, S. Rudd, J. R. D. Taplin and C. Frimodt-Møller, *Biodiversity and Conservation*, 2000, **9**, 33.

52. M. Sosef in "The Biodiversity of African Plants", L. J. G. van der Maesen, X. M. van der Burgt and J. M. van Medenbach de Rooy (eds), Kluwer Academic Publishers, Dordrecht, 1996, 602.

53. M. Sosef, "Glacial rain forest refuges and begonias", Wageningen Agricultural University Papers, 1994, 94–1.

54. M. D. Bengo and J. Maley, *Compt. Rendus Acad. Sci. II*, 1991, **313**, 843.

55. J. Maley in "The Biodiversity of African Plants", L. J. G. van der Maesen, X. M. van der Burgt and J. M. van Medenbach de Rooy (eds), Kluwer Academic Publishers, Dordrecht, 1996, 519.

56. F. White, "The Vegetation of Africa", UNESCO, Paris, 1983.

57. M. Colyn, *Revue Zoologique du Afrique*, 1987, **101**, 183.

58. M. Colyn, A. Gautier-Hion and W. Verheyen, *J. Biogeogr.*, 1991, **18**, 403.

59. A. Prigogine, Aix-en-Provence XIX Congress of Ornithology, 1988, 2537.

60. J. Kingdon, "East African Mammals: an Atlas of Evolution in Africa", Academic Press, London, 1971.

61. S. Bahuchet in "L'alimentation en Fôret Tropicale: Interactions Bioculturelles et Applications au Development", C. M. Hladik, A. Hladik, O. F. Linares, H. Pagezy, A. Semple and M. Hadley (eds), Paris Parthenon/UNESCO, 1993, 37.

62. A. W. Diamond and A. C. Hamilton, *J. Zool.*, 1980, **191**, 379.

63. W. A. Rodgers, C. F. Owens and K. M. Homewood, *J. Biogeogr.*, 1982, **9**, 41.

64. C. A. Brühl, *J. Biogeogr.*, 1997, **24**, 233.

65. P. Tattersfield, *Malacologia*, 1996, **38**, 161.

66. A. C. Hamilton, *Boissiera*, 1975, **24**, 29.

67. J. C. Lovett and I. Friis in "The Biodiversity of African Plants", L. J. G. van der Maesen, X. M. van der Burgt and J. M. van Medenbach de Rooy (eds), Kluwer Academic Publishers, Dordrecht, 1996, 582.

68. M. S. Roy, *Proc. Roy. Soc. Lond. B Biol. Sci.*, 1997, **108**, 1.

69. N. Burgess, C. FitzGibbon and P. Clarke in "East African Ecosystems and Their Conservation", T. R. McClanahan and T. P. Young (eds), Oxford University Press, Oxford, 1996, 329.

70. H. A. C. Eeley, M. J. Lawes and S. E. Piper, *J. Biogeogr.*, 1999, **26**, 595.

71. R. Harmsen, J. R. Spence and W. C. Mahaney, *J. Afr. Earth Sci.*, 1991, **12**, 513.

72. H. A. Osmaston in "Quaternary and Environmental Research on East African Mountains", W. C. Mahaney (ed.), A. A. Balkema, Rotterdam, 1989, 7.

73. W. R. Peltier, *Science*, 1994, **265**, 195.

74. G. H. Denton and T. J. Hughes, "The Last Great Ice Sheets", Wiley, Chichester, 1981.
75. B. Huntley, *J. Biogeogr.*, 1993, **20**, 163.
76. K. D. Bennet, P. C. Tzedakis and K. J. Willis, *J. Biogeogr.*, 1991, **18**, 103.
77. G. H. Denton and T. J. Hughes, "The Last Great Ice Sheets", Wiley, Chichester, 1981.
78. K. J. Willis, *Endeavour*, 1996, **20**, 110.
79. J. C. Roeland, J. Guiot and R. Bonnefille, *Comptes Renous de l'Academie des Sciences Paris. Serie IIMecanique Physique*, 1988, **307**, 1735.
80. R. Bonnefille, J. C. Roeland and J. Guiot, *Nature*, 1990, **346**, 347.
81. A. M. Aucour, C. Hillaire-Marcel and R. Bonnefille, *Geophysical Monographs*, 1993, **78**, 343.
82. J. Maley in "Paleoclimatology and Paleometeorology: Modern and Past Patterns of Global Atmospheric Transport", M. Leinen and M. Sarnthein (eds), Kluwer Academic Publishers, Dordrecht, 1989, 585.
83. R. W. Fairbridge in "Problems in Palaeoclimatology", A. E. M. Nairn (ed.), Wiley-Interscience, London, 1964, 356.
84. H. A. El-Nahhal, *Palaeogeography, Palaeoclimatology, Palaeoecology*, 1994, **100**, 303.
85. L. Dinesen, T. Lehmberg, J. O. Svendsen, L. A. Hansen and J. Fjeldsa, *Ibis*, 1994, **136**, 2.
86. B. Messerli and M. Winiger, *Mt. Res. Dev.*, 1992, **12**, 315.
87. R. Bonnefille and F. Chalié, *Global Planet. Change*, 2000, **26**, 25.
88. J. C. Lovett, *J. Trop. Ecol.*, 1996, **12**, 629.
89. D. S. Wilki and M. C. Trexler, "Central Africa-Global Climate Change and Development", Technical report 1-29 Biodiversity Support Program, 1993.
90. F. Gasse, *Quaternary Sci. Rev.*, 2002, **21**, 737.
91. J. M. Adams, "Global land environments since the last interglacial", Oak Ridge National Laboratory, TN, USA, http://www.esd.ornl.gov/ern/qen/nerc.html., 1997.
92. A. Vincens, D. Schwartz, H. Elenga, I. Ferrera, A. Alexandre, J. Bertaux, A. Mariotti, L. Martin, J. D. Meunier, N. Nguetsop, M. Servant, S. Servant-Vildary and D. Wirrman, *J. Biogeogr.*, 1999, **26**, 879.
93. J. A. Coetzee, *Palaeoecology of Africa*, 1967, **3**, 5.
94. A. C. Hamilton, "Environmental History of East Africa: a Study of the Quaternary", London, Academic Press, 1982.
95. F. A. Street-Perrot and R. A. Perrot in "Global Climates since the Last Glacial Maximum", H. E. Wright, J. E. Kutzbach, T. Webb III, W. F. Ruddiman, F. A. Street-Perrot and P. J. Bartlein (eds), University of Minnesota Press, Minnesota, 1988, 318.
96. D. Jolly, D. M. Taylor, R. A. Marchant, A. C. Hamilton, R. Bonnefille, G. Buchet and G. Riolett, *J. Biogeogr.*, 1997, **24**, 495.
97. E. M. van Zindderen-Bakker and J. A. Coetzee, *Review of Palaeobotany and Palynology*, 1988, **55**, 155.

98. J. Maley in "Paleoclimatology and Paleometeorology: Modern and Past Patterns of Global Atmospheric Transport", M. Leinen and M. Sarnthein (eds), Kluwer Academic Publishers, Dordrecht, 1989, 585.

99. J. Maley and P. Brenac, *Review of Palaeobotany and Palynology*, 1998, **99**, 157.

100. H. Elenga, D. Schwartz and A. Vincens, *Palaeogeography, Palaeoclimatology, Palaeoecology*, 1994, **91**, 345.

101. J. Runge, "New Results of the Late Quaternary Landscape and Vegetation Dynamics in Eastern Zaire (Central Africa)", Gebruder Borntraeger, Berlin, 1995.

102. S. Ning and L. M. Dupont, *Vegetation History and Archaeobotany*, 1997, **6**, 117.

103. F. Marret, J. Scourse, J. H. Jansen and R. Schneider, *Acad. Sci. Paris*, 1999, **329**, 721.

104. S. Ning and L. M. Dupont, *Vegetation History and Archaeobotany*, 1997, **6**, 117.

105. D. Jolly, D. M. Taylor, R. A. Marchant, A. C. Hamilton, R. Bonnefille, G. Buchet and G. Riolett, *J. Biogeogr.*, 1997, **24**, 495.

106. N. A. Sowunmi, *Palaeoecology of Africa*, 1991, **22**, 213.

107. A. Vincens, *Review of Palaeobotany and Palynology*, 1993, **78**, 381.

108. J. M. Maitima, *Quaternary Res.*, 1991, **35**, 234.

109. J. Haffer, *Science*, 1969, **165**, 131.

110. J. Haffer, *Biodiversity and Conservation*, 1997, **6**, 451.

111. P. A. Colinvaux in "Evolution and Environment in Tropical America", B. C. Jackson, A. F. Budd and A. G. Coates (eds), The University of Chicago Press, Chicago, 1996, 359.

112. J. Haffer, *Science*, 1969, **165**, 131.

113. G. T. Prance, "Biological Diversification in the Tropics", Columbia University Press, New York, 1982.

114. J. Haffer, *Biodiversity and Conservation*, 1997, **6**, 451.

115. M. B. Bush, W. D. Gosling and P. A. Colinvaux in "Tropical Rainforest Responses to Climatic Change", M. B. Bush and J. R. Flenley (eds), Springer, Berlin, 2006, 55.

116. R. J. Morley in "Tropical Rainforest Responses to Climatic Change", M. B. Bush and J. R. Flenley (eds), Springer, Berlin, 2006, 1.

117. M. R. Silman in "Tropical Rainforest Responses to Climatic Change", M. B. Bush and J. R. Flenley (eds), Springer, Berlin, 2006, 269.

118. A. P. Kershaw, S. van der Kaars and J. R. Flenley in "Tropical Rainforest Responses to Climatic Change", M. B. Bush and J. R. Flenley (eds), Springer, Berlin, 2006, 77.

119. K. S. Jr. Brown in "Biogeography and Quaternary History in Tropical America", T. C. Whitmore and G. T. Prance (eds), Clarendon Press, Oxford, 1987, 175.

120. J. C. Lovett and I. Friis in "The Biodiversity of African Plants", L. J. G. van der Maesen, X. M. van der Burgt and J. M. van Medenbach de Rooy (eds), Kluwer Academic Publishers, Dordrecht, 1996, 582.

121. J. Fjeldså and J. C. Lovett, *Biodiversity and Conservation*, 1997, **6**, 325.
122. J. C. Lovett, R. Marchant, J. Taplin and W. Küper in "Phylogeny and Conservation", A. Purvis, J. L. Gittleman and T. M. Brooks (eds), Cambridge University Press, Cambridge, 2005, 198.
123. R. Marchant, S. Behera, T. Yamagata, C. Mumbi, *African Journal of Ecology*, 2007, **45**, in press.
124. C. D. Thomas, A. Cameron, R. E. Green, M. Bakkenes, L. J. Beaumont, Y. C. Collingham, B. F. N. Erasmus, M. F. De Siqueira, A. Grainger, L. Hannah, L. Hughes, B. Huntley, A. S. Van Jaarsveld, G. F. Midgley, L. Miles, M. A. Ortega-Huerta, A. Townsend Peterson, O. L. Phillips and S. E. Williams, *Nature*, 2004, **427**, 145.
125. C. J. McClean, J. C. Lovett, W. Küper, L. Hannah, J. H. Sommer, W. Barthlott, M. Termansen, G. F. Smith, S. Tokumine and J. R. D. Taplin, *Ann. MO Bot. Gard.*, 2005, **92**, 139.
126. M. R. Silman in "Tropical Rainforest Responses to Climatic Change", M. B. Bush and J. R. Flenley (eds), Springer, Berlin, 2006, 269.
127. S. D. Torti, P. D. Coley and T. A. Kursar, *Am. Nat.*, 2001, **157**, 141.
128. D. M. Newbery, X. M. van der Burgt and D. M. Newbery, *J. Trop. Ecol.*, 2004, **20**, 131.
129. J. C. Lovett and M. Poudyal, *African Journal of Ecology*, 2006, **44**, 302.
130. W. J. Bond, F. I. Woodward and G. F. Midgley, *New Phytol.*, 2005, **165**, 525.
131. M. Sankaran, N. P. Hanan, R. J. Scholes, J. Ratnam, D. J. Augustine, B. S. Cade, J. Gignoux, S. I. Higgins, X. Le Roux, F. Ludwig, F. Ardo, F. Banyikwa, A. Bronn, G. Bucini, K. K. Caylor, M. B. Coughenour, A. Diouf, W. Ekaya, C. J. Feral, E. C. February, P. G. H. Frost, P. Hiernaux, H. Hrabar, K. L. Metzger, H. H. T. Prins, S. Ringrose, W. Sea, J. Tews, J. Worden and N. Zambatis, *Nature*, 2005, **438**, 846.
132. S. R. James, *Curr. Anthropol.*, 1989, **30**, 1.
133. C. K. Brain and A. Sillent, *Nature*, 1988, **336**, 464.
134. M. I. Bird and J. A. Cali, *Nature*, 1998, **394**, 767.
135. N. Goren-Inbar, N. Alperson, E. K. Mordechai, O. Simchoni, Y. Melamed, A. Ben-Nun and E. Werker, *Science*, 2004, **304**, 725.
136. S. Weiner, Q. Xu, P. Goldberg, J. Liu and B. Ofer, *Science*, 1998, **281**, 251.
137. A. P. Kershaw, S. van der Kaars and J. R. Flenley in "Tropical Rainforest Responses to Climatic Change", M. B. Bush and J. R. Flenley (eds), Springer, Berlin, 2006, 77.
138. A. P. Kershaw, S. van der Kaars and J. R. Flenley in "Tropical Rainforest Responses to Climatic Change", M. B. Bush and J. R. Flenley (eds), Springer, Berlin, 2006, 77.
139. A. P. Kershaw, S. van der Kaars and J. R. Flenley in "Tropical Rainforest Responses to Climatic Change", M. B. Bush and J. R. Flenley (eds), Springer, Berlin, 2006, 77.
140. G. M. Paduano, M. B. Bush, P. A. Baker, S. C. Fritz and G. O. Seltzer, *Palaeogeography, Palaeoclimatology, Palaeoecology*, 2003, **194**, 259.

141. M. B. Bush, J. A. Hanselman and H. Hooghiemstra in "Tropical Rainforest Responses to Climatic Change", M. B. Bush and J. R. Flenley (eds), Springer, Berlin, 2006, 33.

142. J. R. Harlan, "Crops and Man", American Society of Agronomy, Madison, 1992.

143. A. P. Kershaw, S. van der Kaars and J. R. Flenley in "Tropical Rainforest Responses to Climatic Change", M. B. Bush and J. R. Flenley (eds), Springer, Berlin, 2006, 77.

144. M. D. Pohl, K. O. Pope, J. G. Jones, J. S. Jacob, D. R. Piperno, S. D. deFrance, D. L. Lentz, J. A. Gifford, M. E. Danforth and J. K. Josser, *Lat. Am. Antiq.*, 1996, **7**, 355.

145. R. Bonnefille in "Tropical Rainforest Responses to Climatic Change", M. B. Bush and J. R. Flenley (eds), Springer, Berlin, 2006, 117.

146. D. C. Nepstad, A. Verissimo, A. Alencar, C. Nobre, E. Lima, P. Lefebvre, P. Schlesinger, C. Potter, P. Moutinho, E. Mendoza, M. Cochrane and V. Brooks, *Nature*, 1999, **398**, 505.

147. C. H. Cannon, D. R. Peart and M. Leighton, *Science*, 1998, **281**, 1366.

148. P. A. Keddy and C. G. Drummond, *Ecol. Appl.*, 1996, **6**, 748.

149. H. C. Dawkins and M. S. Philip, "Tropical Moist Forest Silviculture and Management: a History of Success and Failure", CAB International, Wallingford, 1998.

150. J. R. Vincent, *Land Econ.*, 1990, **66**, 212.

151. E. B. Barbier and J. C. Burgess, *Land Econ.*, 1997, **73**, 174–195.

152. C. M. Peters, A. H. Gentry and R. Mendelsohn, *Nature*, 1989, **339**, 655.

153. E. F. Bruenig, "Conservation and Management of Tropical Rainforests: An Integrated Approach to Sustainability", CAB International, Wallingford, 1996.

154. R. Godoy, D. Wilkie, H. Overman, A. Cubas, G. Cubas, J. Demmer, K. McSweeney and N. Brokaw, *Nature*, 2000, **406**, 62.

155. J. V. Krutilla, *Am. Econ. Rev.*, 1967, **57**, 777.

156. Earl of Cranbrook and D. S. Edwards, "Belalong, A Tropical Rainforest", The Royal Geographical Society, London, 1994.

157. L. S. Hamilton, J. O. Juvik and F. N. Scatena, "Tropical Montane Cloud Forests", Springer Verlag, Berlin, 1995.

158. S. M. Brooks and T. Spencer, *J. Environ. Manag.*, 1997, **49**, 297.

159. R. A. Kramer, D. D. Richter, S. Pattanayak and N. P. Sharma, *J. Environ. Manag.*, 1997, **49**, 277.

160. N. Myers, R. A. Mittermeier, C. G. Mittermeier, G. A. B. da Fonseca and J. Kent, *Nature*, 2000, **403**, 853.

161. J. C. Lovett, R. Marchant, J. R. D. Taplin and W. Küper in "Phylogeny and Conservation", A. Purvis, J. L. Gittleman and T. M. Brooks (eds), Cambridge University Press, Cambridge, 2005, 198.

162. J. C. Lovett, Mitteilungen aus dem Institut für Allgemeine Botanik Hamburg, 1990, **23a**, 287.

163. J. C. Lovett, *J. Trop. Ecol.*, 1999, **15**, 689.

164. C. J. McClean, J. C. Lovett, W. Küper, L. Hannah, J. H. Sommer, W. Barthlott, M. Termansen, G. F. Smith, S. Tokumine and J. R. D. Taplin, *Ann. MO Bot. Gard.*, 2005, **92**, 139.
165. IPCC, "Climate Change 2001: The Scientific Basis", Cambridge University Press, Cambridge, 2001.
166. IPCC, "Climate Change 2001. Impacts Adaptation and Vulnerability", Cambridge University Press, Cambridge, 2001.
167. L. Hannah, G. F. Midgley, T. Lovejoy, W. J. Bond, M. Bush, J. C. Lovett, D. Scott and F. I. Woodward, *Conservat. Biol.*, 2002, **16**, 264.

The Implementation of International Biodiversity Initiatives: Constraints and Successes

EEVA FURMAN, RIKU VARJOPURO, ROB VAN APELDOORN AND
MIHAI ADAMESCU

1 Towards International Biodiversity Goals

Jared Diamond[1] describes in his book *Collapse* how human societies through-out history have affected their own survival by the way they have used natural resources. The book provides several examples of social behaviour driven by short-term local needs instead of taking a long-term and broad-scale approach to their strategy of welfare but also gives examples in which human behaviour has been sustainable enough to avoid collapse. The sad story of the destruction of civilisation in the Easter Islands, the message of sustainable livelihood of the Nord culture in Iceland and the window to the open future of the Montana community all lead us to think that we do need goals that take a broad perspective and look far into the future, but also raise commitment and understanding by those who are today responsible for their implementation. Goals give a thread to follow for decision making on the strategic level, on the tactical level and finally on the operational level. They are thus important for various scales of societies: for those leading broader entities such as countries, and smaller entities like municipalities, but also as important for those running their everyday life in close connection with biodiversity, *e.g.* farmers, foresters and even urban dwellers.

As the concept of biodiversity was launched as late as the 1980s,[2] early conservation initiatives were developed through other concepts. The romantic age brought the aesthetic value of natural landscapes into high regard. This led to the establishment of national parks. At the change from nineteenth to twentieth century, the establishment of early national parks was based on human needs for valued landscapes and sites and even for moral education.[3,4]

Issues in Environmental Science and Technology, No. 25
Biodiversity Under Threat
Edited by RE Hester and RM Harrison

Many parks still existing today have thus been selected according to their aesthetic value rather than their role in biodiversity conservation, although due to their size and long history as pristine areas they are of a high quality also from the biodiversity point of view. Since early times, many more parks and protected areas have been established, in many cases on the basis of their role in biodiversity conservation. Their establishment has been one of the key conservation means.

The consequences of overharvesting through hunting and fisheries were realised in the mid 1900s which led in the 1980s to several international agreements on species conservation, dealing with threats to species or the habitats they require (see CITES and Ramsar conventions in the next section). Strong support for a species-oriented approach in international conservation policy came from parallel steps taken in science at that time. The emphasis on the species level was maintained until the formulation of the Convention of Biological Diversity in 1992 and, despite its high status, in many cases even beyond that: the conceptual approach to species protection has been maintained until today in national and international conservation policies as one of the key units.

A move towards a more comprehensive and complex approach (ecosystem approach), which takes into consideration entire ecosystems with their complexities, functionalities such as species interactions, and the concept that humans are a part of ecosystems, has recently been observed. This change of vision in goal setting has brought new perspectives upon measuring the success of biodiversity initiatives.

This concept of biodiversity based on the ecosystem approach takes the necessary steps forward, bringing into consideration the concept of sustainability as one very important component of conservation policies. The change in perspective is dramatic, because from conserving biodiversity we are gradually moving towards recognising biodiversity as the foundation of human society development.[5] At present, we are in the middle of the process of understanding biodiversity as having a much more comprehensive value. This approach is imposing tremendous implications upon the effectiveness of conservation strategies and policies, the most obvious of which are those dealing with the integrated approach or those operating across different spatial and temporal scales. In fact, expanding the concept of biodiversity to include human society represents just one step on a long path. The following steps materialise when the move is taken towards understanding our world as one of systems and processes with all its consequences: high dynamics of systems, complexity and uncertainty, time lag, hierarchically structured systems, dissipative structures, open systems, *etc.*[6,7,8]

In this chapter we analyse the evolution of biodiversity governance through the lens of the evolving concept of biodiversity. Measuring constraints and successes of international biodiversity initiatives is a huge task that could not be completed in these few pages and this is not our intention. Rather, our intention is to show how selected international initiatives translate into local practices and to analyse what kinds of constraints have been seen in the implementation and which have led to success in our selected cases.

2 International Initiatives to Set Goals for Biodiversity Conservation

The Convention on Biological Diversity (CBD) was adopted at the 1992 Earth Summit in Rio de Janeiro. The Convention has three main goals: (1) conservation of biological diversity, (2) sustainable use of biodiversity and (3) fair and equitable sharing of benefits arising out of the utilisation of genetic resources. Through the plethora of international initiatives focused on biodiversity, the CBD is having a great impact upon the protection of biodiversity at the global level. Other international initiatives which set goals for biodiversity conservation already existed prior to the CBD. The 1971 Ramsar Convention (or Convention on Wetlands) provides the framework for national and international actions for the conservation and wise use of wetlands and the 1973 CITES Convention (International Trade in Endangered Species of Wild Fauna and Flora) aims to ensure that international trade in specimens of wild animals and plants does not threaten their survival. In addition, a convention concerning the protection of the world's cultural and natural heritage (1972 World Heritage Convention) is a strongly biodiversity-related initiative. The Convention on the Conservation of European Wildlife and Natural Habitats (1979 Bern Convention) deals with special aspects in relation to biodiversity conservation and sustainable use. The Bern Convention aims to conserve especially those species and habitats whose conservation requires co-operation between several governments. The Bonn Convention on Migratory Species (1979) also deals with biodiversity, but concentrates on the terrestrial, marine and avian migratory species.

In addition to those international initiatives which directly target biodiversity, there are several others which have mentioned biodiversity in their secondary goals. The 1994 United Nations Convention to Combat Desertification deals with the processes of land degradation, erosion and human activities such as over-cultivation and deforestation. The 1992 United Nations Framework Convention on Climate Change (UNFCCC) and its instruments, such as the Clean Development Mechanism (CDM), also require biodiversity consideration.[9] The 1997 Kyoto Protocol as a treaty has more powerful and legally binding measures than the UNFCCC. This protocol on combating climate change came into force in 2005. The Cartagena Protocol on Biosafety (2003) aims at protecting biological diversity from the potential risks posed by living modified organisms resulting from modern biotechnology. The Protocol contains reference to a precautionary approach and reaffirms the precautionary language in Principle 15 of the 1992 Rio Declaration on Environment and Development.

3 How are International Goals Implemented?

The CBD, as many other international initiatives, is an outcome of intergovernmental negotiations. Governments of several countries are taking part in the goal setting, implementation and follow up. Decisions are made by the Conference of the Parties (COP) that meets every two years. National delegations, led by governments, include administrators from the ministries of environment and foreign affairs, members of the research community and, in some delegations, representatives of national non-governmental organisations (NGOs). In addition to the various governments, the European Union also has ratified the convention and is a full member of it. Furthermore, there are international NGOs such as IUCN and governmental organisations, *e.g.* UNEP, taking part in the negotiations as observers.

The implementation of the CBD is supported by several regional biodiversity initiatives in various parts of the world, such as the IUCN Regional Biodiversity Programme in Asia (see www.rbp-iucn.lk). Together with regional supra-governmental actors, the United Nations regional programmes on the environment and on the economy have provided the driving force in developing these initiatives. Below we describe the setting for the CBD implementation in Europe.

The Pan-European Biological and Landscape Diversity Strategy (PEBLDS, see www.strategyguide.org/straabou.html) was developed in 1994 by the initiatives of the Council of Europe and the United Nations Environment Programme (UNEP) to support the CBD implementation. In the EU, the European Community Biodiversity Strategy (1998) and its Biodiversity Action Plans (2001) form the main response to the CBD. They cover a wide range of topics in line with the CBD and are centred on the two main EU legal instruments for biodiversity conservation: the EC Birds Directive and the Habitats Directive. There are also European policies which highlight conservation of biodiversity, such as the European Water Framework Directive (WFD) aiming at improving the state of polluted waters and ensuring the sustainable use of water resources in surface, coastal and groundwater, all of these influencing biodiversity. The implementation will take place according to locally developed River Basin Management Plans by using ecological criteria.[10]

The revised European Common Agricultural Policy (CAP) and its instruments, such as the new European Agricultural Fund for Rural development, will affect large areas and their potential for effective implementation of the CBD.[11] The integration of nature management, landscape and environmental concerns into the CAP has gained momentum with the CAP reforms adopted in June 2003.[12] The agreements in the WTO are set in a different forum, but in reality they do bind the EU and therefore can have considerable influence on the CAP and hence on biodiversity.[13] For marine and aquatic biodiversity, the Common Fishery Policy (CFP) of the EU balances between direct sectoral interests and the ecological goals, *i.e.* between overexploitation and sustainable marine ecosystems.[14]

In 2003, the EU's Biodiversity Strategy implementation was reviewed and debated during the International Stakeholder conference on biodiversity in Malahide, Ireland. The outcome of this conference includes the Message from Malahide, which sets more concrete targets and priority actions to help achieve the 2010 target.[15] The next International Stakeholder conference will take place in Stavanger, Norway, in the spring of 2007.

Next we discuss how the implementation of the international goals has been planned to operate. Under the CBD, "Each Contracting Party shall, in accordance with its particular conditions and capabilities:

(a) Develop national strategies, plans or programmes for the conservation and sustainable use of biological diversity or adapt for this purpose existing strategies, plans or programmes which shall reflect, *inter alia*, the measures set out in this Convention relevant to the Contracting Party concerned; and
(b) Integrate, as far as possible and as appropriate, the conservation and sustainable use of biological diversity into relevant sectoral or cross-sectoral plans, programmes and policies" (CBD, Article 6).

Regional and supra-national action plans help neighbouring countries to strive together to reach global goals. The regional support in Europe was described above. There are shared regional goals, action plans and guidelines as well as regional workshops to support countries to gain institutional capabilities to deal with the challenges of national implementation. Right now, reaching the EU 2010 target to halt the loss of biodiversity in Europe is high on the agenda in science and environmental politics as the deadline is getting closer. For example, the EU has launched the concept of Countdown 2010 to promote action.

On the national level, the implementation could be described as four-phased. The process starts by setting the responsibilities and agreeing upon the documents needed for planning and developing the national implementation procedure. In the second phase, the deliberation of national goal setting takes place. The third phase incorporates and translates national biodiversity goals into national policy. Finally, enforcement relies on the institutional capabilities and support from co-operating stakeholders.[16] When looking at those involved in the implementation, the actors, one soon realises that roles are found at all hierarchical levels from global, regional, national to local. The roles of the actors vary. International and regional actors support and catalyse the implementation processes on the national level. National-level actors translate the international goals to be implemented at lower levels, while local-level organisational actors mediate the goals to those carrying out local practices and, finally, local actors perform the practices that affect success in reaching the international goals.

The obligations of the CBD are general in character, thus being difficult to transpose into specific legislation. In some countries, CBD has formed the basis for building national legislation for conservation. For those countries where conservation legislation existed prior to the CBD, the implementation of the CBD required a careful analysis of its correspondence with existing legislation to find the need for alterations. Even in countries with well-developed conservation policy, some amendments have been needed, in particular in legislation on the use of natural resources. On the other hand, many of the obligations of the CBD do not require regulation and can be fulfilled through other policy instruments. The national strategies set requirements for various sectors and administrative levels. Action plans have been developed to enhance the practical application of a National Strategy. Next we provide two examples of national implementation. Sections 3.1 and 3.2 describe how the governments of the UK and Finland, respectively, have tackled the challenge.

3.1   How has Finland Organised the Implementation of International Biodiversity Goals?

Finland has committed itself to various international initiatives including the CBD, Ramsar, CITES and Bonn conventions. Membership of the EU necessitates that Finland implements its directives, important from the biodiversity perspective being the Habitat Directive and the Birds Directive. The European Conservation Strategy of the EU also sets the framework for biodiversity management in Finland. These international-level policies are both implemented into the legislation as well as applied in practice. The active parties at the national level are Parliament and the various Ministries. Each Ministry is responsible for the conservation and sustainable use of biodiversity within its field of jurisdiction, as well as for making proposals for measures promoting biodiversity (Council of State Decision dated December 21, 1995). The National Biodiversity Action Plan (BAP) ensures that international legal obligations are fulfilled. Together with the National Biodiversity Strategy, the BAP stresses the importance of cross-ministerial collaboration in reaching the goals. This is important, *e.g.* in agreeing upon the development of economic incentives, environmental labelling and certification, environmental impact assessment at the policy level, environmental management and dissemination of information. The regional level includes public bodies such as regional environmental centres, regional council, forestry boards, road districts and rural area business districts, all of which have specific responsibilities in the management of biodiversity. These bodies have responsibilities for ensuring that nature conservation planning, environmental impact assessments, regional and forestry planning, as well as the formation of regional development plans, are in line with the BAP. Local-level authorities have been given a special role in the BAP. They can enhance its implementation through land-use planning, in master plans, in making official decisions, in monitoring the state of biodiversity through nature inventories, in implementing environmentally related projects and by raising awareness among citizens and local actors. Research institutes and universities, both publicly funded, have the responsibility of providing increasing knowledge through research and monitoring activities. Knowledge on the state of biodiversity and its threats, as well as on the effectiveness of various management measures, is expected to be forwarded by the research community. Information on biodiversity is synthesised and disseminated through the electronic information exchange centre (see www.environment.fi/ lumonet), although there has been criticism that not enough information is available beyond the administration, especially when dealing with marine areas.

Although many of the existing institutions, including policies and organisational responsibilities, have a long history and their practices have taken a certain form along the years, the development of the National Strategy and Action Plan aims for a new form of governance. New interactive preparation procedures have opened the policy-making to other sectors than the environmental one, to non-governmental organisations and to interest groups at all levels of management. However, policy-making is still based on representation and local participants are included in setting objectives for management only to a limited degree.[17]

3.2   How has the UK Organised the Implementation of International Biodiversity Goals?

In the United Kingdom, the CBD is implemented through a national biodiversity action plan which includes a country study and a strategy (BAP). It was drawn up in 1994 after the UK had ratified the convention. The action plan divides institutions and practices into three issues: the conservation of habitats, conservation outside natural habitats, and sustainable use. In addition, it highlights the role of partnerships, education and data management, as well as the UK support for conservation of biodiversity overseas.

As in Finland, the UK has committed itself to many other international initiatives that relate to biodiversity in addition to the CBD and is implementing the EU directives. The UK has a long history in conservation. For example, the Protection of Wild Birds Act takes us to 1880. However, the 1981 Wildlife and Countryside Act and the Nature Conservation and Amenity Lands Order from 1985 are the bedrock of nature conservation in today's UK and they enable the country to meet its international obligations when supported by additional regulation arising from regular reviewing. The Government has taken the lead in preparing the action plan. However, a wide range of sectors of society, including some 300 organisations, were consulted and a seminar organised to bring together views from government and four biodiversity-oriented statutory agencies (English Nature, Scottish Natural Heritage, the Countryside Council of Wales and the Environment Service of the Department of Environment for Northern Ireland), as well as from academia and the NGOs.

The government is formally committed to carrying out reviews of wildlife on a regular basis, taking action accordingly and reviewing legislation. The four statutory agencies in the field of biodiversity contribute to biodiversity management through the Joint Nature Conservation Committee. The agencies advise central and local governments on policies, carry out habitat protection and encourage enjoyment of the countryside. In addition to the government and the agencies, there is a long tradition in the UK for voluntary movements which play an important role in owning and managing nature reserves and other areas of wildlife. These movements include the National Trust and the Royal Society for the Protection of Birds. The government enables biodiversity research through various departments such as the National Environment Research Council (NERC) and the Economic and Social Research Council. They have committed themselves to specific requirements of the BAP. The BAP of the UK emphasises the role of landowners, occupiers and managers, planners, regulatory bodies and conservation organisations, as well as those who set the economic and political framework.[18]

3.3 Alternative Routes from International Goals to Local Level Practices: Ranomafana National Park in Madagascar as a Case Study

In most cases, the implementation of international goals takes the planned route via supra-national and national level to the local level. This is not, however, always the case. We know how markets in a foreign country may successfully put pressure on practices at local level that go beyond the obligations set by the country in question. For example, a group of Finnish paper companies signed a moratorium for selected Russian Karelian forests in the 1990s.[19] Examples can be found from other areas of environmental innovations as well, as described by Mickwitz *et al.*[20]

Implementing international biodiversity goals at local level takes place through international or foreign actors who see local biodiversity as valuable internationally and are prepared to use economic incentives to support its conservation. Many international funding agencies have strict environmental, including biodiversity, standards which they require from projects which they fund[21,22] (see also http://www.worldbank.org/). However, another type of result of foreign intervention has taken place in Madagascar.

With its unique and highly endemic fauna and flora, Madagascar has attracted conservation interest both nationally and internationally. Madagascar is also one of the poorest countries in the world. Therefore, in 1988, the government of Madagascar decided to tackle the dual needs of conservation and development by enforcing a 15-year national environmental action plan. This activated significant international donors such as the World Bank and the US Agency for International Development to take part in funding the implementation. The key activities of the action include running Integrated Conservation and Development Projects (ICDPs) and establishing the Malagasy national parks association.

Ranomafana national park, with its long history of foreign research activities, was included in the action plan that was officially established in 1991.[23] The link between foreign and local participants for twenty years has led to a thriving collaboration where local communities select foreign-funded ICDP projects worth 50 000 dollars annually and, in addition, gain 100 jobs in tourism. A national participant, the ValBio research and training centre, collaborates with foreign and local actors and is highly praised by the local communities.[24] Much has been gained in the field of conservation infrastructure and knowledge. The majority of local people in the peripheral zones see conservation as a valuable goal and they are happy that the park was created. Change of local practices towards a more conservation-supportive direction has taken place. It has, however, been claimed that the likelihood of the practices moving back to the previous, unsustainable forms is high if economic support is stopped: the locals simply cannot afford to select practices on any other criterion but the economic one.[25]

4   When Implementation is Being Constrained

Interventions seldom work perfectly as planned. Constraints to effective implementation vary from technical lack of information or means to political and cultural issues when the rationale of nature conservation is contested. Table 1 illustrates four main categories of constraint as identified and further discussed elsewhere by Furman *et al.*[26]

Analyses of conflicts during implementation and management show a wide range of constraints but also provide activities and key factors that play an important role in reaching the formulated goals. Table 2 includes some of the factors that support the implementation of controversial policies. These were analysed in 12 case studies involving biodiversity conflict management in Europe.

The factors in Tables 1 and 2 illustrate that the implementation of policies is a dynamic process with specific characteristics and conditions to which attention should be paid.

First, policy implementation is a dynamic and iterative process of decision making involving communication, interaction and coordination. This can take place at a single policy level or several different levels. Together they form multi-level biodiversity governance. Second, policy implementation is a multi-level decision-making process consisting of phases which are not necessarily consecutive. These include the problem definition phase, the phase of generating alternative solutions and their possible trade-offs, the phase of producing action plans and programmes and implementing them and, finally, the phase of

Table 1   Four categories of policy constraints.[27]

Policy constraints	
A. Social constraints	– process (participation, communication, legitimacy, rights, relationships, values, ethics, perceptions and attitudes)
B. Policy constraints	– knowledge and information (*e.g.* data and research)
	– policy options
	– acceptance and legitimacy
	– integration and coordination
C. Economic constraints	– capital
	– labour
	– natural resources (land, water and others)
D. Resource constraints	– rarity of habitat, species, ecosystem characteristics

Table 2 Success factors in biodiversity management.[28]

- early identification of possible conflicts/risk analysis
- stakeholder involvement (consultation and participation)
- collaboration
- building trust and awareness raising (learning process) between stakeholders
- role of key persons
- building legitimacy (of groups, arguments)
- long-term plans for the use of a resource (*e.g.* area) based on sustainability
- different policy strategies and variation in policy measures or technical instruments
- integrated planning/plan
- interdisciplinary research
- community-based integrated management plans
- increased stakeholder knowledge
- communication and information processes
- sufficient resources (*e.g.* for compensation payments, research and data collection) to avoid frustration
- capacity building of local and national organisations

evaluation.[29] Third, policy implementation takes place in the context of politics and administration, both of which vary in space and time. Fourth, policy implementation is a decision-making process that, regardless of its level or phase, is characterised by a strong dependency between the development of the policy content, the policy network and the kind of process by which rules of negotiation and co-operation are determined.[30,31] Last of all, policy implementation can include integration of different (sectoral) policies as is illustrated at the highest policy level of Europe. European policies (*e.g.* CAP, CFP) and even many of the EU directives force member states increasingly to integrate sectoral policies with each other to support sustainable use of landscapes and biodiversity (*e.g.* forests and marine ecosystems; compare also CBD). This integration does, however, differ between the policy levels: the demand is highest at the lower national administration levels (regional, local) where administrators and policy makers have to integrate policies coming from various other levels (national, European).

5   Does Implementation Lead to Wanted Results?

Many biodiversity policies aim at changing the state of biodiversity itself in terms of species, habitats or system components (with CBD and the Birds and Habitat Directives as the most recent intercontinental and European examples; see also Section 1). For this reason, the simplest way to measure success in implementing policies is to look at the changes in the object of the initiative, that is the state of biodiversity and changes in its status. This is called goal attainment[32] or environmental/ecological effectiveness.[33]

This kind of measurement of policy effectiveness makes the assumption that relations between actors at one policy level and between different policy levels are more or less linear and do not change much in time. However, this is not the case when, for example, higher levels are directly related to lower levels excluding the levels in between.[33] Besides, good communication and coordination (within and between policy levels) are assumed to be the most important conditions that are required for success in reaching the goals. Given these conditions, certain issues are further assumed:[34]

- all participants agree about the content of the discourse and know what they are talking about and why,
- the participants have balanced and clear relationships; they agree with the existing distribution of power and means or resources, information is shared, there is co-operation, and coalitions are clear, as is the role that each participant plays and his/her position,
- all participants agree with the rules of the communication; for example, who is allowed to participate, whether we work by consensus or not, if law is respected, legitimacy and rights are respected and, finally,
- there is enough time for the implementation process.

In traditional hierarchical top-down implementation processes in less complex decision-making situations, such as between a single ministry and its lower administrative level, these assumptions will be agreed and they will result in an effective implementation of policy. An example could be action that needs to be taken against an alien species. However, today most policy implementation processes involve a wide variety of participants, which causes changes in the content, relationships and rules of sharing power and responsibilities between policy levels. Bad communication and coordination lead to problems or constraints or even conflicts. This is typically illustrated by the following features: the content of the discourse becomes unclear or debatable within or between levels, visions and attitudes are no longer accepted, some elements in the discourse start to prevail, relationships become unbalanced, cooperation stops; strong coalitions become important; participants try to influence the distribution of means (money, information, law rules); roles of participants change (from peacemaker to troublemaker, from stage manager to negotiator); the rules change (from consensus to the judge/court); and time becomes limited for the implementation process. One example is the conservation of a species or its habitat, the conservation of which would affect other livelihoods such as fisheries (*e.g.* seal protection) or forestry (*e.g.* conservation of habitat in decaying trees). These possible constraints and conflicts illustrate why many implementation processes are not efficient, as already described in Tables 1 and 2. They demonstrate that the measurement of effectiveness of policy implementation by using only the change in the state of biodiversity and its status as criteria is too simple.

This kind of measurement of policy effectiveness has severe drawbacks. Due to lack of a broader set of criteria for measuring effectiveness, the causes of the unintended trend are not known, nor are the factors that influence the state of biodiversity. Therefore it is difficult to make changes to the policy.

Indicators and reports which reflect the success of the national implementation of the CBD are collected and synthesised by the CBD secretariat and reported to the Subsidiary Body for Scientific, Technical and Technological Advice (SBSTTA), which gives recommendations for the COP (see Table 3).[35] The indicators have, however, been under much debate ever since the convention was initiated and new lists are drawn up by various bodies, which typically compile the indicators separately for each sector. Indicators on the state of biodiversity dominate the lists, although recently the drivers and pressures that cause changes in biodiversity also have gained recognition (Table 3). The response indicators are striking in their absence, partly due to the lack of monitoring systems. The national reports that are sent to the CBD secretariat on a regular basis are also required to contain a section on the processes through which the reports are developed. Different countries have solved this question in different ways. On the whole, stakeholders from various levels and sectors, and with various roles in society, were involved as planned.

There are many criteria that could be used to improve the analyses of the effectiveness of policy implementation along with the state of biodiversity. Criteria applied by Hildén *et al.*[36] in their evaluation of environmental policy instruments could be used as a basis. They propose ten criteria: relevance, impact, environmental effectiveness, cost-effectiveness, applicability, transparency and participatory rights, equity, flexibility, predictability and sustainability. When framing the scope of the evaluation, the DPSIR approach (drivers, pressures, state, impact and response) provides a model for a broader picture of the complex problem (see Figure 1). It also takes into account that the effectiveness of biodiversity and biodiversity policies do depend not only on conservation policies but on policies in other sectors as well. The DPSIR concept, however, still lacks the input of discourses on its implementation. The criteria which reflect local participants, their practices and the discourses which arise from their everyday life are also needed.[37]

For these reasons, Young[38] describes an analysis of institutional and environmental effectiveness of international policy implementation. For institutional effectiveness, Young refers to the institutional level where all (inter)national policies and programmes start.

One reason for targeting the measurement of institutional effectiveness is the complexity of the implementation process and the dynamics at all levels of decision making or negotiation (compare also ref. 16). Another reason is the need to integrate sectoral policies as highlighted earlier in this chapter. This requires coordination between different levels as well as social acceptance, *i.e.* legitimacy for the policy at all levels. This last aspect raises the importance of the cost-effectiveness of policies and stresses the need for a shared understanding about the policies in hand. This is enhanced through investment in participation and communication.[39]

6   Mediation to Help Reach Effectiveness and Legitimacy

We have shown above that the traditional way of tackling policy processes and practices linked with them as a linear conceive-decide-implement continuum

Table 3 Provisional CBD Indicators for policy evaluation included in the Malahide process.[35]

A: Focal area	B: Indicator for immediate testing	C: Possible indicators for development by SBSTTA or Working Groups
Status and trends of the components of biological diversity	Trends in extent of selected biomes, ecosystems and habitats	
	Trends in abundance and distribution of selected species	Change in status of threatened species (Red List indicator under development) Trends in genetic diversity of domesticated animals, cultivated plants and fish species of major socioeconomic importance
	Coverage of protected areas	
Sustainable use		Area of forest, agricultural and aquaculture ecosystems under sustainable management Proportion of products derived from sustainable sources
Threats to biodiversity	Nitrogen deposition	Numbers and cost of alien invasions
Ecosystem integrity and ecosystem goods and services	Marine trophic index	Application to freshwater and possibly other ecosystems Connectivity/fragmentation of ecosystems Incidence of human-induced ecosystem failure Health and well-being of people living in biodiversity-based-resource dependent communities
	Water quality in aquatic ecosystems	Biodiversity used in food and medicine
Status of traditional knowledge, innovations and practices	Status and trends of linguistic diversity and numbers of speakers of indigenous languages	Further indicators to be identified by WG-8j
Status of access and benefit-sharing		Indicator to be identified by WG-ABS
Status of resource transfers	Official development assistance provided in support of the Convention (OECD-DAC-Statistics Committee)	Indicator for technology transfer

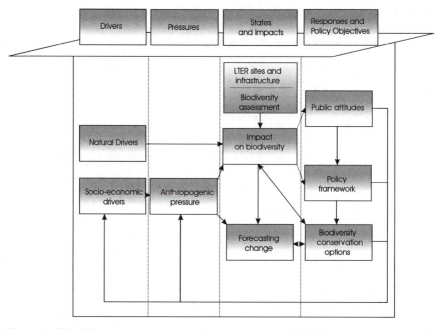

Figure 1   The drivers, pressures, state, impact, response (DPSIR)-concept was initially
introduced by the European environment agency (EEA) and is today broadly
used in policy and research institutions such as the Research Network of
Excellence, the ALTER-Net (taken from ALTER-Net Newsletter Number 2,
www.alter-net.info).

from top to bottom can lead to problems, especially when the local dynamics
have been neglected. In this predetermined continuum, which is often the way
nature conservation policies operate, the goal for the policy is defined at a high
level while its execution is delegated to lower levels of administration. This far,
decisions and delegation take place inside the spheres of policy making and
administration and things run more or less smoothly. In the actual implementation, the defined goal has to leave the sphere of bureaucracy and step into
"the real world" as a form of policy measure that should reach the expected
results in conserving biodiversity. However, because the goal was defined in
terms of the conservation policy itself, it does not necessarily resonate with the
social, cultural and/or economic context of the implementation, assuming that
the scientific basis of the decision was correct and suits well the ecological
context of the implementation. Therefore, complications in the implementation
may emerge, as described for the Natura 2000 implementation in France,[40] or
in implementation of a new nature conservation scheme in The Netherlands.[41]
In both cases the processes prepared at national level had to be changed and
rethought in the phase of implementation at local level.

    The problem presented here becomes reality only occasionally, but is possible
in any implementation of a top-down practice of policy. It becomes perceptible
when it is recognised that policy implementation can function as a constitutive

of politics[41] and it is not only about execution of political decisions. The traditional top-down practice of policy places "the politics" at the top level, as the stage at which the actual decisions are finalised. This makes sense only in the context of top-down policy practice in the sphere of bureaucracy but it is, in fact, based on a misconception of the policy process.

In most cases, the conservation measures are meant to change people's use of natural resources or other human activities that affect biodiversity. However, in cases where complications in the implementation emerge, the problem lies in the discrepancy between policy goal (or its translation to a policy measure) and the local practices. The local practices, *e.g.* everyday logistics of a fishing community or a single farmer, are based on a complex interplay of social, cultural and/or economic needs and capabilities with ecological resources and, furthermore, embedded in large-scale socio-economic processes. In addition, they have a time span that seldom coincides with that set for implementing the biodiversity goals. Because of the interplay of the various existing needs into which the conservation need is now brought, the conservation goal has to be debated against the social, cultural and economic dynamics. In other words, "the politics" is ignited again and this is something that the top-down management regime is not able to handle, because the assumption is that the political debates preceded the decision. The local natural-resource managers, who have the duty of implementing and enforcing the decisions in a top-down regime, have to deal with the new political debates at the local level. The dilemma is that the top-down regime does not give the local managers a mandate for such debates, nor require qualifications for such.

Where to look for the solution to this problem? How to reconcile integration of more functions at wider scales of space and time and in the meantime stimulate participation of individual local people? What are the capabilities of local stakeholders to take part in the policy and politics and how could social learning be supported? In a policy environment where there are various needs that contradict or support each other, one needs an integrated approach that can explore the potential conflicts and synergies between the issues. The need for integration of policies was recognised already in the 1980s and is endorsed in many policy initiatives and regulations.[42]

Regarding policy-making for conservation, this means that the social, cultural and economic aspects are taken into account together with the ecological ones. At the high-level of politics, discussions and debates are abstract and there the integration can be endorsed as a principle or in a more concrete way; for example, by defining a set of criteria for concrete decision making at the lower levels. On the other hand, at the level of implementation, the principle of integration has to find concrete application. For instance, impact assessments need to include criteria on social, cultural and economic aspects, against which the benefits and the costs of projects are assessed. In addition, aspects should be considered from a long-term perspective. The integrated management approach that has been applied, for instance, in coastal areas and in landscape management, provides good examples of linking local-level politics to higher-level goal setting. In other words, decision-making on natural resources requires a landscape/ecosystem approach, in which conservation goals are considered together with other region-specific goals. It is important that the region, its goals and priorities, are taken as the starting point when incorporating policy goals coming from above. Let us look at one example of this, which comes from Romania and is presented in Section 6.1.

6.1 Braila Islands: Starting Management from Local Socio-environmental Needs Before Linking with International Goals

The small Romanian island of Braila is the most important remnant of the former Inner Delta of the Danube River. The Inner Delta was a large landscape dominated by islands and floodplain wetlands, located upstream from the Coastal Danube Delta, from which 80% of the surface area was converted for agricultural purposes after 1960. Complex multi-disciplinary research activities have been carried out for almost 15 years in Braila Island, focused on ecosystem complexes and socio-economic systems. As a result, it was recognised that this area is important for habitat and species conservation according to the EU Habitat and Bird directives. This extensive and intensive investigation made it possible to identify the structure, function and carrying capacity of this complex of ecosystems. At the same time, the high dynamics of the ecosystem complex as a consequence of natural and anthropogenic drivers and pressures were recognised.

The main conflicts identified were those between local communities and the state-owned agencies, mainly due to "export" of goods (like wood and fish) and limited access to resources. On the other hand, people have been using the entire area, including Braila, for centuries, developing their society without posing a real threat to the system. Based on these arguments and taking into consideration also the principles of the CBD Ecosystem Approach and the need to comply with the international agreements like CBD and European directives, a first version of the integrated management plan was developed. For more than two years this version of the management plan was discussed and improved, with local communities seeking both understanding and negotiations so that all involved stakeholders could agree upon the objectives of the management plan. After reaching consensus at this level of debate, the achievements were examined by involving other decision-making levels, starting with the immediate top administrative level, *i.e.* county level, then proposing this site as protected at national (Natural Park) and international level (Ramsar site, SPA). Having a scientific base for the development of the management plan was a good solution for both conservation and sustainable use.

International agreements (either at European or global level) came later as a justification in negotiations, especially at county and national level, but the real change came from understanding that the local nature forms the basis for the socio-economic development of the area. Not surprisingly, the local people with a long tradition in directly using goods and services provided by different ecosystems knew this very well.

In practice, integration means negotiating between different goals and criteria for development. It involves political debate and deliberation that can be done in various ways and through different processes. Integrative discussions are needed at various levels, but the issues discussed and their degree of concreteness differ from one level to another. Regarding a solution to the problem of policy implementation across levels depicted above, the crucial level seems to be the landscape one, because at this level issues discussed are concrete enough to show the actual synergies and conflicts between different goals. At this level, the conservation issues become meaningful to people. It has been noted that often when experts and policy makers try to highlight the environmental risks to increase environmental awareness they talk about global-level risks. Hajer and Versteeg[43] argue that "the storyline of 'global nature' in particular would lack the connection with everyday life and thereby have disempowering effect". This emphasises that integration at the high levels is a necessary but not a sufficient condition. The social and economic issues may have been discussed and their integration to biodiversity policy been solved in the political discussions and debates at the high levels preceding the political decision. However, when policy implementation is seen as an aspect of politics, the integration has to be planned and discussed again at lower levels. The following Section 6.2 depicts a progression from intense conflict to constructive solutions. Much time was, however, needed for carrying out the entire process.

6.2   From Hunger Strikes to Voluntary Measures: Forest Conservation in Karvia

The first proposal on sites in Finland to be included in Natura 2000 was published in 1997. The proposal caused resistance across Finland; the environment authorities received approximately 14 000 letters of complaint. Landowners considered that their opinions were not listened to when doing the preparation and the plans did not take into consideration social and cultural aspects. In the municipality of Karvia in south-west Finland, protest was exceptionally strong. Four landowners went on a hunger strike to protest against the proposed Natura 2000 network, which ended after one week when both the minister of agriculture and forestry and the minister of the environment visited the scene and nearly half of the designated areas in Karvia were withdrawn from the Natura 2000 proposal. The planning and implementation of Natura 2000 proved to be unsatisfactory and relationships between main participants had turned hostile. Simultaneously with Natura 2000 a province-wide nature conservation enquiry was completed, which proposed 300 new conservation areas in the province of Satakunta, where Karvia is situated. The organisation responsible for planning was aware that the landowners were indignant over Natura 2000 and had to proceed carefully and constructively since it was obliged to integrate private owners of forest into forest conservation planning. However, no organisation was officially responsible for coordinating the interaction between planning organisations and landowners.

Today, ten years after the Natura 2000 process started, there is a new forest conservation model in the region. Now landowners are voluntarily offering their areas for conservation against financial compensation. Karvia region is one of the pilot areas for this new voluntarily based "natural value trade". The model was elaborated and connected as a part of the forest biodiversity programme for southern Finland (METSO) and is now spreading to other parts of Finland as well. How did the region's forest conservation evolve from hunger strikes to pilot project of voluntary, collaborative activities? The need to proceed carefully with landowners and the lack of clarity in the coordination duties caused waste of human and organisational resources. The process required several years of active co-operation resulting in improved confidence, knowledge and understanding of the operational model. Without intention, this difficult situation led to interaction and reorganisation. The modification resulted as part of the broader social development dynamics. This course of events can be illustrated through an adaptive cycle model.[44] The cycle comprises four phases: (1) conservation, (2) release or creative destruction, (3) reorganisation and (4) exploitation or growth and control. In the case of Karvia the phase of creative destruction (2) occurred when legislation about forestry and nature conservation, as well as discussion about socially, economically and ecologically sustainable development, had just attained a mature phase (1). The conflict raised by landowners broke the fragile consensus and co-operation and was followed by a conflict which reorganised the regional forest conservation field, which in turn led to a new political invention, natural value trade (3). As a political innovation it is growing as part of a larger initiative "Forest biodiversity programme for Southern Finland (METSO)", which continues to assemble economic and human capital to support the innovation (4).[45,46]

6.3 Turning Symbolic Participation into Effective Deliberation

Integrated management, as it is commonly understood, implies participatory decision-making and planning.[47,48] Public participation has been promoted strongly and also practised a lot in environmental decision-making. One important reason for this has been the complexity of environmental issues: complexity in terms of knowledge needs, but also regarding the interests vested in the decisions to be made. Public participation has been seen to deliver a broader, more representative range of knowledge and values to bear on the complexity and uncertainty of the environmental problems. It has become a cornerstone of many institutional approaches to sustainability.[49]

However, carrying out public participation has proved to be difficult from the perspective of set expectations.[50,51] Organising participatory processes is rather common, but meeting the high expectations is not that frequent. On some occasions participation has been seen to be only symbolic when authorities arrange hearings or committees without real intent to allow it to affect their decisions, while in other cases the participatory process has not brought shared views and has not led to any conclusions. Hajer[51] has observed, though, that "it is not so much participation itself that is the problem but the very conditions under which the exchange of ideas has to take place". Many of the problems come down to the notion that in participatory processes it is different "stake-holders" or "interest groups" that are represented. This is, of course, based on a basic democratic principle, but its consequence is that individuals or groups of people are reduced to preferences and interests[52,49] – in the worst case to one preference or one interest per group. A deliberative process, in which people are represented in their reduced forms, places people against each other: this form of participation does not encourage open communication or mind sets to seek for shared views. When the participants are from the outset positioned against each other, or in certain assumed coalitions according to their interest, the deliberative process is structured in such a way that its potential for producing new ideas or new coalitions and even new practices is lost. In other words, "performative" aspects of deliberative processes that could produce new ways of acting together and understanding the problems at hand are not utilised.

Avoiding a participatory process that produces oppositional positions requires more attention to the settings in which deliberation takes place. The setting of policy making, *e.g.* how deliberation is arranged, where it takes place and which means of communication are used, has often been neglected while attention has been paid to content of argumentation or results of voting. Hajer[51] clearly shows that the setting itself influences "what is said, what can be said, and what can be said with influence".

It is important to be sensitive to the outcomes of the different arrangements set for the deliberative events. Official working groups may serve as effective channels for collecting information and values relevant to making conservation policies, but they have a very strong bias effect due to the skewed representativeness. Therefore, alternative forms of deliberation should be utilised to support the process. For instance, Hajer[51] analysed one environmental decision-making process in which he identified that the process evolved in very typical stages from draft plans prepared by experts and authorities via public protests to renewed and more acceptable plans. However, in the process there was an unconventional occurrence of artistic happening that influenced the plans and the region's future in general. In this happening, the same participants of the official planning process were placed in new roles and the organisational responsibility was taken by someone other than the authorities. This step opened up the setting for new ways of communication and raised fresh ideas for the official process. The fixed statutory-based modes of participation seldom allow escape from predetermined positions and therefore novel turns are needed to give space for fresh solutions. One process where governance is broadly shared is the farmer's role in biodiversity management in the broad countryside. Empowering processes can thus be very innovative. Let us now turn to community-led biodiversity management in the Netherlands.

6.4   A Successful Community-led Biodiversity Management Process: the Farming for Nature Initiative

In some parts of the Netherlands where nature values are high, farmers that want to manage their farms in a nature-friendly way can get compensation money from the Ministry of Agriculture, Nature and Food Quality (ANF). But nature values also exist outside these areas. The ministry has no financial instruments for co-operation with farmers to manage this aspect of nature. For these areas a new, non-governmental initiative has been taken, called "Farming for Nature" (FfN). Although the initiative is non-governmental, the funding is mainly public.

The main goal of FfN is to stimulate farmers to manage nature and the landscape. Using a contract, they are obliged to manage their land in a landscape- and nature-friendly way to maintain important ecological processes and conditions. The process is done by guaranteeing financial compensation for ecologically valuable services from a regional fund. Farmers sign a long-term contract under private law that ensures nature development and a fair remuneration for farmers. They are free to decide which management form suits their circumstances best and which parts of their farms will be used for landscape or new landscape elements or nature management. The farmer is still the owner of the land and enters into an agreement on the way the land is to be used and this is laid down in a contract (in the form of a servitude or lease). Developing nature and landscape takes time. Therefore, continuous management is needed and long-term agreements are drawn up. The farmer has to agree to maintain the ecologically valuable landscape elements and the same holds for his successor. In the case of a nature-oriented farm, the land can only be sold to someone who practises nature-oriented farming, so the desired management form is continued.

A significant part of the money in the fund is provided by regional, mainly governmental and administrative bodies, which are responsible for practical implementation of biodiversity and other environmental goals. As stated, the type of farming is not result-oriented but condition-oriented. However, monitoring and evaluation of the chosen management has to be described in the contract. It has not yet been decided who will do the monitoring – the farmer or, for example, some volunteers. The remuneration money is not based on precisely defined "target species or plant communities" or predescribed measures, but rather on consulting with the farmer on various landscape elements and the most appropriate operating plan. So the farmer is still in charge of his own farm.[53,54]

As pointed out in this chapter, local protests over implementation of policies infer that policy implementation can lead to new political debates and, furthermore, the typical statutory-based arrangements for participation may not serve as an arena for the political debates as they easily rule out the unofficial, non-expert participants. An emphasis on representational statements in such processes does not allow utilisation of the local resource-use practices and behaviours in decision making[49] because "public understanding of human and non-human aspects of local and other contexts are opaque to the categories of 'preferences' and 'values'". Representational participation reduces the content of decision making and obscures how everyday practices and behaviours might be engaged. This breaks the link from conservation policy to ecosystem since "ultimately it is these practices and behaviours that perform the outcomes."[49] The Farming for Nature initiative described in this chapter shows one example of how to organise communication in a way that resonates with local practices in using natural resources. The scheme is negotiated through the farmers' practices and emphasis is put on practices that provide suitable conditions for biodiversity rather than setting specific goals in terms of species or habitats.

7 Conclusions

The way to involve the public is to ensure the long-term engagement of conservation into practical everyday context. This requires that management of natural resources is arranged as part of an area-oriented landscape management approach in which nature conservation is taken as one important aspect among economic, social and cultural issues. The management practices should also include new and innovative arrangements for deliberation to avoid the conditions that set participants against each other. Problems and protests in implementation cannot be avoided – after all, implementation can be constitutive of politics – but more flexible management practices can better cope with such emerging problems. Ideally, local politics could in the long run reflect back to the international level, either directly or by using regional and national pathways. This way, international goals could open up to flexibility, learn from local experiences and tune their contents to match the needs of the future.

Acknowledgements

We would like to thank Ms Hanne Nurminen for kindly helping with the collection of material for this article and Joosef Valli for help in technical editing. The study was supported by the European Union within the FP 6 Network of Excellence ALTER-Net (Project no. GOCE-CT-2003-505298).

References

1. J. Diamond, "Collapse. How Societies Choose to Fail or Succeed", Viking, New York, 2005.
2. E. O. Wilson and F. M. Peters (eds), "Biodiversity", National Academy Press, Washington, D.C., 1988.
3. S. Mirams, *Aust. Hist. Stud.*, 2002, **33**, 249.
4. S. Schama, "Landscape and Memory", Harper Collins Publishers, London, 1995.

5. A. Vadineanu, "Dezvoltarea durabila" Vol I, Editura Universitatii din Bucuresti, 1998.
6. N. Botnariuc, "Conceptia si metoda sistemica in biologia generala", Edit, Acad. Romane, Bucuresti, 1967.
7. H. T. Odum, "Ecological and General Systems: An Introduction to Systems Ecology", University Press of Colorado, Niwot, CO., USA, 1994.
8. A. Vadineanu, "Sustainable development: Theory and Practice regarding the Transition of Socio-Economic Systems towards Sustainability" UNESCO-CEPES Studies on Science and Culture, Bucuresti, 2001.
9. K. Brown, W. N. Adger, E. Boyd, E. Corbera-Elizalde and S. Shackley, "How do CDM projects contribute to sustainable development?", Technical report 16, Tyndall Centre for Climate Change Research, University of East Anglia, Norwich, UK, 2004.
10. P. G. Holland, *Flow Meas. Instrum.*, 2002, **13**, 277.
11. P. F. Donald, G. Pisano, M. D. Rayment and D. J. Pain, *Agr. Ecosyst. Environ.*, 2002, **89**, 167.
12. F. M. Brouwer and F. E. Godeschalk, Nature Management, Landscape and the CAP. LEI Report 3.0401. LEI, The Hague. 2003.
13. H. Nillson, *Journal of Cleaner Production*, 2004, **123**, 460.
14. C. Iglesias-Malvido, D. Garza-Gil and M. Varela-Lafuente, *Marine Policy*, 2002, **26**, 403.
15. G. Duke (ed.), "Biodiversity and the EU – Sustaining Lives, Sustaining Livelihoods, Conference report", Department of the Environment, Heritage and Local Government, Ireland, 2005.
16. I. M. Bouwma, C. A. Balduk and F. M. Brouwer, "Conventions – convenience or constraint? A framework for analysing the implementation of international nature conservation conventions and programmes", Alterra-report 727, Wageningen University, 2003.
17. P. Kangas, J. Pekka Jäppinen, M. von Weissenberg and H. Karjalainen (eds), "National Action Plan for Biodiversity in Finland, 1997–2005", The Finnish Ministry of the Environment, 1998.
18. "Biodiversity: the UK action plan", HMSO Publications Centre, London, 1994.
19. Greenpeace, "Buying destruction. A Greenpeace report for corporate consumers of forest products", 1999.
20. P. Mickwitz , H. Hyvättinen and P. Kivimaa, *Journal of Cleaner Production*, in press.
21. OECD, "Integrating the Rio Conventions into Development Co-operation. The DAC Guidelines", Paris, 2002.
22. OECD, "Good Practice Guidance on Applying Strategic Environmental Assessment (Sea) in Development Co-operation", DCD/DAC(2006)37, Paris, 2006.
23. P. -C. Wright in "Making Parks Work for Preserving Tropical Nature", J. Terborgh, C. van Schaik, L. Davenport and M. Rao (eds), Island Press, 2002, 279.
24. T. E. Lovejoy, *Trends Ecol. Evol.*, 2006, **21**(6), 329.

25. R. R. Marcus, *Hum. Ecol.*, 2001, **29**(4), 381.
26. E. Furman, R. van Apeldoorn, B. Petriccione, T. Gottsberger, A. Gaaff and E. Primmer, "Means to measure success of implementing (inter)national conservation strategies: successes and constraints", Report WPR4-2005-02 of the project ALTERNet (www.alter-net.info), 2005.
27. E. Hellström, "Conflict Cultures – Qualitative Comparative Analysis of Environmental Conflicts in Forestry", Silva Fennica Monographs 2, 2001.
28. P. S. Jones, J. Young and A. D. Watt (eds), "Biodiversity conflict management, a report of the BIOFORM project", 2005.
29. P. Glück, A. C. Mendes, and I. Neven (eds), "Making NFPS work: supporting factors and procedural aspects", Publication Series of the Institute of Forest Sector Policy and Economics, Vol. 48, 53, 2003.
30. I. Neven, Proceedings of the seminar of Cost Action E19 "National Forest programmes in a European Context", Vienna, 2003.
31. E. Turnhout, "Ecological indicators in Dutch nature conservation. Science and policy intertwined in the classification and evaluation of nature", Vrije Universiteit, Amsterdam, 2003.
32. G. Kütting, "Environment, Society and International Relations. Towards More Effective International Agreements", Routledge, 2000.
33. M. Roman and E. Vedung, Paper presented at the American Evaluation Society Conference, Honolulu, 2000.
34. I. Neven and S. Schruijer, Proceedings of MOPAN Conference Coalitions & Conclusions, Tilburg University, Netherlands, 2005.
35. CBD (Convention on Biological Diversity), "Strategic Plan: future evaluation of progress", UNEP/CBD/COP/VII/30, 2004.
36. M. Hildén, J. Lepola, P. Mickwitz, A. Mulders, M. Palosaari, J. Similä, S. Sjöblom and E. Vedung, "Evaluation of environmental policy instruments – a case study of the Finnish pulp & paper and chemical industries", *Monographs of the Boreal Environment Research*, 2002, **21**, 1.
37. H. Svarstad, D. Rothman, F. Wätzold, L. K. Petersen and H. Siepel, "Land Use Policy", in press.
38. O. R. Young, "International Governance. Protecting the Environment in a Stateless Society", Cornell University Press, 1992.
39. E. Primmer and S. Kyllönen, *Forest Pol. Econ.*, 2006, **8**(8), 838.
40. P. Alphandéry and A. Fortier, *Sociologia Ruralis*, 2001, **41**, 311.
41. M. Hajer in "Deliberative Policy Analysis. Understanding Governance in the Network Society", M. Hajer and H. Wagenaar, eds., Cambridge University Press, UK, 2003, p. 88.
42. P. Mickwitz, "Environmental Policy Evaluation. Concepts and Practice", Commentationes Scientiarum Socialium 66, The Finnish Society of Sciences and Letters, Helsinki, 2006.
43. M. Hajer and W. Versteeg, *J. Environ. Pol. Plann.*, 2005, **7**, 175.
44. H. L. Gunderson and C. S. Holling, "Panarchy: Understanding Transformations in Systems of Humans and Nature", Island Press, London, 2001.
45. J. Hiedanpää, *Landsc. Urban Plann.*, 2002, **61**, 113.

46. J. Hiedanpää in "Uusi metsäkirja", R. Jalonen, I. Hanski, T. Kuuluvainen, E. Nikinmaa, P. Puttonen, K. Raitio and O. Tahvonen (eds), Gaudeamus Press, Finland, 2006, 377.

47. J. Bellamy, G. McDonald, G. Syme and J. Butterworth, *Soc. Nat. Resour.*, 1999, **12**, 337.

48. R. Kay and J. Alder, "Coastal planning and management", E. & F. N. Spon, London, 1999.

49. S. Healy, *J. Environ. Pol. Plann.*, 2005, **7**, 239.

50. D. Bloomfield, K. Collins, C. Fry and R. Munton, *Environ. Plann. C*, 2001, **19**, 501.

51. M. Hajer, *Admin. Soc.*, 2005, **36**, 624.

52. J. E. Innes and D. E. Booher, "Public Participation in Planning. New Strategies for the 21st Century", University of California, Institute of Urban and Regional Planning, Berkeley, CA, USA, 2000.

53. M. Buizer, T. Ekamper, A. van de Berg, R. Kwak and C. de Vries, "Verhalen van Biesland. Boeren voor Natuur", Wageningen University, Alterra, 2005.

54. A. H. F. Stortelder, R. A. M. Schrijver, H. Alberts, A. van den Berg, R. G. M Kwak, K. R. de Poel, J. H. J. Schaminée, I. M. van den Top and P. A. M. Visschedijk, "Boeren voor natuur; de slechtste grond is de beste", Wageningen University, Alterra, 2001.

Biodiversity Assessment and Change – the Challenge of Appropriate Methods

MICHAEL BREDEMEIER, PETER DENNIS, NORBERT SAUBERER, BRUNO PETRICCIONE, KATALIN TÖRÖK, CRISTIANA COCCIUFA, GIUSEPPE MORABITO AND ALESSANDRA PUGNETTI

1 Introduction

1.1 The Progressive Inclusion of Biodiversity Measures in Environmental Monitoring

The recognition of the importance of monitoring within ecosystems emerged only since the mid twentieth century. The concept of biological indicators, as opposed to particular target "headline" organisms and the measurement of these alongside broader environmental parameters, was adopted in ecosystem monitoring with the establishment of the United Nations Environment Programme.[1] A formal recommendation to focus on biological diversity in biological monitoring appeared in the Brundtland Report.[2] There followed widespread acceptance that the quality of air, water and soil can be monitored far more effectively with the use of indicator species than by environmental monitoring of chemical pollutants or climate alone.[3] Early emphases of European monitoring programmes sought to gauge the state of marine fisheries under increasing harvesting, and forest health as affected by acid deposition, but this soon developed into surveillance of particular plant and animal species, where the conservation of biological diversity became a priority objective in certain European countries as concern mounted over habitat loss and declines in species.[3,4] The CORINE Biotopes Programme was the first pan-European assessment of biotopes of major importance for nature conservation.[5] The essential purpose of long-term monitoring was advocated in the UNEP Global Biodiversity Assessment, that such monitoring was critical "to identify human-made changes from natural changes".[6]

Issues in Environmental Science and Technology, No. 25
Biodiversity Under Threat
Edited by RE Hester and RM Harrison
© The Royal Society of Chemistry, 2007

1.2   The Challenge of Adequately Representing Complexity

Biodiversity comprises the expression of life on earth in all its various forms and at all its relevant levels of complexity, in a hierarchy from genes to the biosphere. This richness, however, is not easily grasped and expressed in operational measures. We need scientifically well-based approaches to represent the inherent complexity and still adequately describe the diversity of life and the services provided for humanity.

In this chapter we describe the methodological challenge related to biodiversity research and monitoring. Beyond the introduction, we evaluate surrogate measures, indicator species and indices of diversity. Surrogates are used to characterise biodiversity where direct measurement of biological diversity is not feasible, whereas indicator species and diversity indices act to summarise information on biodiversity and can serve to communicate it. Terminological distinction is not always straightforward and clear, as the section will show.

Next we consider how these measures of biodiversity are best structured to identify the effect of environmental factors, namely a Long-Term Ecological Research (LTER) network, illustrated using case studies from Italy. Long-Term Ecological Research sites are equipped to continuously monitor environmental and biological variables enabling interdisciplinary experimental research. The research should provide comprehensive information on the state of biodiversity and how it responds to environmental fluctuations and trends. The comprehensive data should allow an assessment of ecosystem functions and the connection with ecosystem services for humanity. Long-Term Ecological Research sites cannot realistically represent the broad range of ecosystems represented across Europe and we consider the importance of additional monitoring in the overall landscape or "wider countryside" for gauging the status of biodiversity in a later section.

The fundamental task of communicating knowledge about the composition and state of biodiversity, and how it is measured, to the policy and public arenas is dealt with in the final section. Key stakeholder groups and the public at large must understand the observed changes in biodiversity and the future consequences for ecosystem function and the quality of human life in order to appreciate the purpose of environmental monitoring.

1.3   Complexity and Ambiguity of the Term Biodiversity

"Overall richness of life on earth" is an appealing linguistic expression, apt to convey the thought of a great natural heritage and the need for its conservation, but how to connect it to measurable quantities? Our knowledge about the "richness of life" on earth is seen to be dramatically incomplete when considering some key numbers. The total number of existing species is unknown, estimates range from 2.5 to 30 million and more.[7,8] The number of actually recorded and described species is currently at 1.5–1.75 million and continuously increasing, owing to intensive taxonomic research. Hence, anything from about

5 to 60% of all species has been recorded by man, and there is no way of knowing the exact figure. Concern about an ongoing erosion and decline of biodiversity is hence based on facts other than precisely known total numbers: it is the apparent loss of some "charismatic" key species (*e.g.* tiger and gorilla) in particular that attracts attention and raises concern, as well as the apparent loss and degradation of many habitats in general. The presumed (and indeed very plausible) link of habitat loss and species loss is a strong line of reasoning in the biodiversity debate.

1.3.1 Hierarchical Level Model. One important consideration is that bio-diversity is expressed at different organisational and spatial scales, which are hierarchically nested, and each of these scales can be an important level in its quantification. The hierarchical levels extend from bio-molecules to the bio-sphere, comprising genes, populations, communities, ecosystems, habitats, bio-geographical regions, biomes and finally the entire biosphere. The hierarchical approach can contribute to make quantifications of biodiversity more feasible and more precise in their meaning. For instance, quantification and comparison of populations with respect to their total gene pool or the richness of variation at certain key gene loci are becoming possible with the help of recent modern molecular genetic techniques.[9] Diversity of traits in microfauna or microbial populations, which could never be studied on the basis of counts and descriptions at the level of individuals, can efficiently be discerned by employing molecular markers. Likewise, on the large spatial scale, remote-sensing-based assessments of the change of habitat extensions and habitat quality can enable qualified appraisals of related changes in biological diversity.

1.4 Approaches to Reduce Complexity

There are several meaningful concepts and tools to handle the complexity inherent in biological diversity. Many of the established parameters date back in time much longer than biodiversity science, since diversity in general can be treated as a problem of information theory (amount of "diverse" information in a signal) and has been addressed with measures such as the Shannon Index long before biodiversity emerged as a topic in science.

The Shannon Index can be regarded as the classic in the field of diversity indicators and as the basis for the development of a multitude of further indices which are commonly used to address different facets of the numerical expression of biodiversity. The formula for the Shannon Index is:

$$S = -\Sigma(p_i \log(p_i)) \qquad \text{with } p_i = n_i/N \qquad (1)$$

N is the total number of individuals in a sample, n_i are the individuals of species I and A is the total number of species observed. However, this expression continues to increase along with A, the total species number.

$$S_{max} = \log(A); \text{ (even distribution of all species assumed in this case)} \qquad (2)$$

To normalise it, the proportion of S to the log of all species is calculated

$$E = S/\log (A) \qquad (3)$$

This normalised term is called the "Evenness" of species distribution in a sample. An evenness value of 1 would hence indicate maximum evenness of the species represented in a sample, lower values indicating an increased dominance of a few species. However, it cannot be assumed that the highest observed evenness values always indicate the ecologically most favourable or sustainable conditions![7,10]

It is further important to note that the Shannon Index and Evenness only give numerical information on the sample considered and no information concerning distribution in space. However, species counts plus the Shannon Index and/or Evenness together provide a complete and quantitative set of information on how many species (of one or more taxa) there are in a sample (or habitat), and how the abundances of the those species are distributed relative to each other (see Figure 1).

Another approach to handle the complexity associated with biodiversity and the different levels on which it is expressed is the distinction of α-, β- and γ-diversity. α-diversity designates diversity within a habitat, β-diversity is the measure within a mosaic of habitats, including borderline effects, and γ-diversity refers to the diversity within a bio-geographical region or country.[11] This classification reflects the integrated and hierarchically nested spatial scales on which biodiversity becomes manifest. It has frequently been supplemented by further classes, designated as δ-, ε-diversity and so on, addressing still different situations and problems of quantitative biodiversity assessments.

1.4.1 Species Numbers and Abundances. Although a number of more-or-less sophisticated indices to quantify biological diversity are available, an exploration of the literature quickly reveals that in most instances simple species numbers and abundances are the most popular measures of diversity in

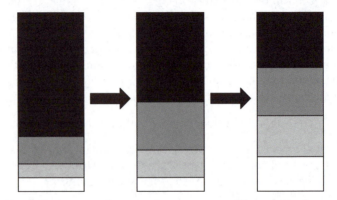

Figure 1 Increasing evenness between the occurrence of four species in a sample.

practical use. This is surely resulting from the fact that species counts and abundance assessments are the very basic and intuitive way to address bio-diversity-related questions. Counting and estimating would naturally be the first thing to do when exploring the biological diversity within a habitat and, in fact, most reporting is based on this simple and straightforward approach. Indicators of trend founded on combined species, like the common farmland bird index[12] or the Pan-European Species Trend Index (STI)[13] are commonly used in large-scale reporting schemes. They represent merely an aggregation of single observations of species numbers and abundances, but are still more frequently used than the sophisticated diversity indices of higher mathematical and logical complexity.

1.4.2 Development of Surrogates, Indicators and Aggregated Indices. Having identified that the reduction of complexity is a major problem to solve, we have to consider in which ways it could best be achieved for our purposes. Well-chosen and evident surrogates, indicators and indices, which are scientifically well founded and broadly accepted by the pertinent scientific community as well as stakeholders, political actors and decision makers, would indeed be a great accomplishment. In the next section, we focus on this challenge.

2 Surrogate Measures, Indicators and Indices

It is no secret that scientists are in pursuit of a general, unifying model to explain the universe. In this instance, it is biodiversity rather than the broader universe, but it is nonetheless extremely complex and widely unknown. Thor-ough studies of, for instance, the species richness of a limited area requires an intensive effort by many collaborators, using much time and money. Despite the immense effort, it is likely that many inconspicuous organisms of obscure taxa would be overlooked due to the general lack of taxonomic expertise. Additionally, the urgency to formulate conservation measures in these decades of rapid biodiversity loss demands the use of short-cuts in biodiversity assess-ment. Altogether, this has led to the development of indicators, indices or so-called surrogates of biodiversity.

2.1 Forerunners and First Steps in the 1980s

One can argue that some studies carried out in the 1960s are forerunners of the biodiversity indicator approach.[14,15] While searching for factors that determine species diversity, they explored the importance of environmental spatial hetero-geneity for species richness patterns. These studies and others were motivated by ecological theory. However, in the 1980s conservation biologists started to explore this topic. In the woodlands of Lincolnshire it was found that rare plants act as an umbrella to include the occurrence of 99% of all other plants.[16] Murphy and Wilcox[17] explored the usefulness of a vertebrate-based conservation management for an insect group. They compared data on species richness on three geographic scales, partly included vascular plants in their analysis, and found reasonable support for their approach. Nevertheless, there are few quotable studies in the 1980s, but new concepts and catchwords were established at the end of that decade.

2.2   A Muddle of Terms

Besides the more general terms of biodiversity indicator or biodiversity surrogate some more focused concepts emerged. For instance, the concept of an umbrella taxon, which is defined as a "species with large area requirements, which, if given sufficient protected habitat area, will bring many other species under protection".[18,19] Or the term flagship species, *i.e.* a limited set of umbrella species chosen for their charismatic appearance (*e.g.* tiger, rhino or elephant). Many more terms and concepts are under discussion (*e.g.* focal and target taxon, keystone species, cross-taxon congruence) and this muddle of terms is frequently criticised and has yet to be clarified.[20–24] There is some consensus that the term surrogate is the most general one. Some researchers deny this and say that the concept of indicator is good enough. However, until a taxon has been demonstrated to be an indicator/surrogate it must remain a potential indicator/surrogate.[25] Mostly used in materials science originally, the word surrogate basically means a substitute. It quickly gained popularity since 1995, although already used as early as 1988 by Landres *et al.*[20] ("the indicator is a surrogate measure"). Altogether, the number of studies on biodiversity indicators/surrogates boomed since the middle 1990s and, anticipating the summary below partly, the philosopher's stone has not yet been found. Effectively, some indicators have been detected and developed that are suitable to represent some aspects of biodiversity, at the most.

2.3   "It Starts with the Right Question" or "the Choice of Values and Measures"

The motivation to use one of the various biodiversity surrogates and measures is manifold. In agricultural landscapes, for instance, we perhaps need biodiversity indicators for biological control, ecosystem functioning or nature conservation.[26] Each mentioned aspect requires at least one indicator and the quality and reliability of potential indicators must be tested rigorously. Nevertheless, the output in the domain of biodiversity indicator research is dominated by its appliance in conservation practice. Again, what indicator would best suit our purpose? Do we need a biodiversity indicator to locate an area of raised species richness or to find the best solution for a representative reserve system?[27] Or, are we searching for indicators that allow decision makers to set measures for the maintenance of biodiversity? Or, a conservation practitioner looks for an indicator to evaluate the success or failure of restoration measures that have been carried out. To demonstrate the rapid development that took place in the last years it makes sense to focus on one practical example.

2.3.1 Deadwood as a Showcase. Forests in Europe have been heavily exploited for many centuries. Logs are harvested long before the trees reach their natural span of life. In an average managed forest only a small amount of deadwood remains. Saproxylic species (*i.e.* species that demand dead and decaying wood) suffered seriously due to habitat loss and fragmentation.[28,29,30] However, about a third of the European forest species need deadwood and veteran trees to some extent for their survival.[31] In Germany, 25% of the beetle fauna depend on deadwood[32] and in Britain 1800 invertebrate species are saproxylic.[33] So, the establishment of conservation sites for saproxylic species became a main issue. Additionally, discussions and recommendations focus on increasing the amount of deadwood in commercially managed forests. Immediately it is apparent that it is time consuming and costly to assess the whole array of saproxylic species. Consequently, potential indicators and surrogates were explored[34,35,36,37] and the cost-effectiveness of surrogates was evaluated.[38]

There is now ample evidence that raising levels of deadwood in forests will promote abundance and richness of deadwood-demanding species. So, why not measure the amount of deadwood as a surrogate for saproxylic species richness? This was proven to be pretty successful for several taxa and it is cost-efficient when data collection is integrated in the national forest inventories.[38] On the other hand, if there isn't any nationally organised forest inventory it might be more efficient to concentrate on some deadwood-demanding umbrella species like woodpeckers.[35,39] However, for the overall biodiversity, dead and decaying wood is not a comprehensive indicator,[38] because other factors on the landscape- and stand-scale also play an important role. Another problem is that deadwood threshold values are poorly understood and identified only for a few species.[40,41] Altogether, in the case of the saproxylic species diversity in Europe, some steps for a profound and cost-effective biodiversity surrogate have been taken, but more needs to be done.

2.4 Adoption of the Biodiversity Surrogate Approach

Some hundred studies on biodiversity surrogates with real field data from around the world have been published in the last two decades. The outcome is ambiguous and no clear picture has emerged so far. What we do know is that the scale of analysis is critical. Scale is recognised as an important issue in ecology,[42] well documented for species richness patterns (*e.g.* Rahbek[43]), and also the effectiveness of biodiversity indicators is markedly influenced.[44,45] On average the predictive power of biodiversity indicators increases from local to global scale (*e.g.* Pharo and Beattie[46] compared to Ricketts *et al.*[47]). Furthermore, it seems that surrogates developed in one geographical region or within a special habitat type are not easily transferable into another context.[45] Reyers and van Jaarsveld[48] demonstrated that the assessment techniques used also have a strong influence on the effectiveness of biodiversity surrogates. The quality of the input data used and observation bias are at least partly a black box, and they both could have serious effects on the outcome of the analysis.

Despite these limitations some progress can be observed. So, environmental surrogates can improve how well species in a reserve system are represented.[49,50] Empirical evidence grows that some taxa could be used as surrogates for a broader aspect of biodiversity. It has been shown, for instance, that vascular plants or birds are efficient indicators at least in some contexts and at certain scales.[51,52,53] Another promising method is the so-called multi-taxa approach: while single groups may fail to serve as biodiversity surrogates, a selection of a set of taxa with different ecological requirements might circumvent this problem.[47,53] This approach directly leads to the development of biodiversity indices, which are increasingly used in biodiversity monitoring.

2.5 Biodiversity Indices

Recently, especially in the light of the world-wide aim to reduce the current rate of biodiversity loss significantly by 2010,[54] the variety of biodiversity indices is growing. Already applied in some countries (*e.g.* Britain, Switzerland, The Netherlands), biodiversity indices increasingly become important on a European[55,56] and world-wide scale.[57] If kept simple, they are smart enough to influence decision makers and they are valuable tools for communication with the public.

In the next section, we examine in detail which indicators of biodiversity should be employed at intensively investigated LTER sites, which are particularly important ecological research objects around the world.

3 Indicators of Biodiversity and its Change for LTER Sites

3.1 LTER Sites and Biodiversity: Basic Concepts and Keystones

Long-Term Ecological Research (LTER) sites consist of various monitoring and research facilities, creating a network across the world. Their data can be profitably used to address research questions on several environmental topics. The concept of LTER implies: (a) long data series; (b) data on ecosystem traits and processes; and (c) a shift from monitoring activities, performed on a regular basis, to research activities. The International Long-Term Ecological Research (ILTER) network was founded in 1993 by the United States of America and had 34 contributing countries by 2006, reflecting the increased appreciation of the importance of long-term research in assessing and resolving complex environmental issues.[58] This was a means to meet a growing need for communication and collaboration among long-term ecological researchers and to provide a scientific forum to encourage data-sharing at a local, regional and national level, co-operation on global projects, and to integrate findings and deliver sound, peer-reviewed research to decision makers and the public. The ILTER network is not the only LTER programme but it is the only one with: (a) a global network of research sites in a wide array of ecosystems worldwide; (b) a focus on long-term site-based research; (c) a governance structure and research mandates built on a "bottom-up" rather than a "top-down" approach.[58]

The usefulness of the ILTER is readily illustrated by recent research to investigate the response of forests to hurricane and typhoon events along the Pacific and Atlantic coasts in the USA (presentation by Steven Hamburg, head of US LTER, Brown University, Rhode Island, USA). Data on the location, period, power category, average power/year from LTER forest sites experiencing tropical cyclones in the USA, Japan and Taiwan revealed that forests responded differently to these storms in the Old and New World.

The following environmental trends were identified by ILTER and external scientists:[58] (a) an accelerating increase in sea temperature and glacial melt as a consequence of climate change with implications for ecosystems, societies and human health; (b) ILTER contributes to sustainable development, especially in the areas of human health, education and access to natural resources linked to environmental conditions, through the participation of developing countries where water scarcity, lack of electricity and environmentally induced diseases prevail; (c) biodiversity loss: biodiversity is one of the key natural resources on which human societies rely. Particularly in developing countries, biodiversity services are essential to humans, assuring survival support by agricultural production, water supply and quality, and pest control; on the other side, in developed countries, pressures on biodiversity are of great concern due to population density and over-exploited landscapes and natural resources. Modifications in biodiversity status and processes can be complicated and slow, affected by several factors, thus requiring long-term and multi-disciplinary approaches, features that ILTER can easily provide.

By 2006, LTER Europe included the Czech Republic, Hungary, Israel, Latvia, Lithuania, Poland, Romania, Slovenia, Slovakia, Ukraine, Austria, France, Italy, Germany, Switzerland and the United Kingdom. A biodiversity gradient can be discerned across Europe, increasing approximately from north to south and from east to west. Formal LTER sites with geo-referenced data are located in the United Kingdom, Switzerland and Eastern Europe[59] because of their political-historical association with ILTER from its inception. If the total number of LTER-like facilities are considered, this widens the distribution of research sites to include Spain, Portugal and Greece. Further countries could contribute data collected in long-term-based research sites to the Network. Hence, biodiversity research topics are already strongly addressed at the European level. A major goal of ALTER-Net[60] under the Integration Objective I3 (a network of long-term multi-functional, inter-disciplinary ecosystem research sites) is to integrate long-term ecosystem research and increase capacity at a national level.

3.2 LTER Networks at Pan-European Level in Practice: the Experience of UN-ECE CLRTAP ICP IM and ICP Forests

Whereas a dedicated pan-European network for biodiversity monitoring and research is still lacking, large and continuous monitoring schemes implementing surveys on biodiversity status and change have been in operation at a pan-European level since 1985 under the UN-ECE Convention on Long-range Trans-boundary Air Pollution (CLRTAP), in particular for the forest ecosystem. The International Co-operative Programmes (ICPs) identify air pollution effects on the environment (including biodiversity) through monitoring, modelling and scientific review.[61] The International Cooperative Programme on Assessment and Monitoring of Air Pollution Effects on Forests (ICP Forests) collects data in close co-operation with the European Commission (under Regulation no. 2152/2003 *Forest Focus*) and determines cause–effect relationships of changes in forests due to air pollution and other stresses by means of large-scale monitoring. The International Cooperative Programme on Integrated Monitoring of Air Pollution Effects on Ecosystems (ICP IM) determines and predicts the state of ecosystems or river catchments and their changes from a long-term perspective with respect to the regional variation and impact of air pollutants. Both ICPs have been co-operating closely for a number of years now, although the objects under study appear different. ICP IM is focusing on catchments in undisturbed ecosystems while ICP Forests monitors both unmanaged and regularly managed forest ecosystems. Most ICP IM sites are, however, located within forest areas and many countries have linked their plots of both programmes within one monitoring system. One result of this intensive co-operation is the harmonisation of assessment methods. Detailed information on the programmes is available on the web sites of UN-ECE ICP[62] and UN-ECE ICP IM.[63]

The activities of ICP IM began as a joint Nordic co-operation programme under the Nordic Council of Ministers in the mid 1980s. From 1989 to 1991 it was run as a pilot programme under the CLRTAP and became a permanent ICP in 1993. The main objectives of the ICP IM are: (a) monitoring of the biological, chemical and physical state of ecosystems (catchments/plots) over time in order to provide an explanation of changes caused by environmental factors, including natural changes, air pollution and climate change, with the aim to provide a scientific basis for emission control; (b) development and validation of models for the simulation of ecosystem responses in order to estimate responses to actual or predicted changes in pollution stress, and make regional assessments in concert with survey data; (c) bio-monitoring to detect natural changes, in particular to assess effects of air pollutants and climate change. The full implementation of the ICP IM will allow determining ecological effects of heavy metals, persistent organic substances and tropospheric ozone. A primary concern is the provision of scientific and statistically reliable data that can be used in modelling and decision making. The ICP IM sites (mostly forested catchments) are located in undisturbed areas, such as natural parks. The ICP IM network presently covers about 50 sites, with on-going data submission, in 21 countries.

ICP Forests was established under CLRTAP in 1986. In 1987 the European Commission (EC) also started to monitor forest condition in the EU Member States. However, ICP Forests and the EC merged their previous two monitoring programmes into a joint one in 1991. Since then, both have been monitoring forest condition and publishing their results jointly. Consequently most of the activities of ICP Forests are carried out in close co-operation with the EC. ICP Forests pursues the following mandate: (a) to monitor effects of anthropogenic stress factors (in particular air pollution) and natural stress factors on the condition and development of forest ecosystems in Europe; (b) to contribute to a better understanding of cause–effect relationships in forest ecosystem functioning in various parts of Europe. For each part of the mandate ICP Forests has implemented a separate monitoring intensity level. At Level I the large scale variation of forest condition is assessed by means of an extensive survey on more than 6000 plots. At Level II intensive monitoring is carried out on 800 plots in 30 countries in order to trace in detail the influence of specific stress factors in main forest ecosystem types. On these plots a larger number of key factors is measured. Apart from air pollution, ICP Forests has widened the scope of its programme to the topics of biodiversity and climate change. In view of these topics, the major objectives of the intensive monitoring at Level II are, in particular, the assessment of: (a) responses of forest ecosystems to air pollution and its changes; (b) differences between present loads and critical loads of atmospheric deposition (tolerable long-term inputs in order to protect the sustainability of the ecosystems); (c) impacts of atmospheric deposition on the ecosystem condition according to scenario analyses; (d) changes in carbon storage in forests (net carbon sequestration); (e) changes in indicators related to the various functions of forest ecosystems to assess long-term sustainability. Both parts of the programme – extensive monitoring on Level I and intensive monitoring on Level II – yield the potential to transfer process information gained on the plot-level (Level II) to the European scale (Level I). The methods for the assessment of the chemical, physical and biological parameters are harmonised throughout both ICPs and are laid down in two manuals (available on the web sites given in refs 62 and 63).

Surveys carried out in both programmes[61] are crown condition, foliar chemistry, species composition of the ground vegetation, soil solid-phase chemistry, soil solution chemistry, tree growth, atmospheric deposition, meteorology, phenology and litter-fall. Surveys that are carried out at the ICP Forests Level II only are ozone injury and remote sensing, whereas soil biology, surface-water chemistry and bird inventories are assessed only by ICP Integrated Monitoring. The number of parameters assessed within the surveys is large; some of them are mandatory, others are optional. The responsibility for selecting the sample plots for each survey and for choosing optional parameters lies with the National Focal Centres (NFCs).

Ground vegetation has been assessed on ICP Forests Level II plots since the 1990s as a biological indicator for effects of chemical deposition, and these make a core contribution to biodiversity monitoring. Under Regulation (EC) no. 2152/2003 Forest Focus, and in line with the ICP Forests strategy, new projects and developments have been initiated recently with the aim of contributing to monitoring some aspects of forest biodiversity. The pilot project "Forest Biodiversity Test-phase Assessments (ForestBIOTA)" aims to harmonise monitoring methods for the assessment of stand structure, deadwood and for lichens growing on tree bark (epiphytic lichens). An ecological classification of forests has also been implemented. The new methods were successfully tested on 107 plots located in 12 European countries in the period 2004–2005. The results of the ForestBIOTA project support the EU demonstration project "BioSoil" that will be carried out on a larger number of Level I plots (6000) in the period 2006–2007. Links to the national forest inventories of many countries are also being intensified in order to provide reliable and comparable information on the biological diversity of European forests.[64]

3.3 Biodiversity Status and Change in Forest Ecosystems: Examples from Italy

Italy entered the International Long-Term Ecological Research (ILTER) network in 2006 after a core group of scientists implemented the Network in the late 1990s. The network of sites was realised through the wider participation of researchers and scientists from public agencies, universities and research institutes coordinated by a National Steering Committee. The National Forest Service – CONECOFOR Office (Corpo Forestale dello Stato – Ufficio CONECOFOR) – has been a partner in the Italian LTER from its instigation. The National Steering Committee proposed sites and research activities during 2005. In 2006, a priority list of Italian sites was produced from which ten suitable sites were finally selected after revision of the scientific methods by external experts. These sites have been grouped into five "macro-sites", each representing a particular ecosystem (forest, freshwater, marine environment, Antarctica Ocean, Himalayan lakes). Sites are linked to each other by ecologically and biogeographically similar traits and may include more than one research station.

3.3.1 Forest Monitoring Activities within LTER-Italy. The "Forest" macro-site includes three sites: (a) site 01 "Forests of the Alps", where the main biotic communities are primary/secondary spruce (*Picea abies*)-dominated forests; (b) site 02 "Forests of the Apennines", mainly *Fagus sylvatica* old-growth forests and mixed coppice stands with secondary meadows; (c) site 03 "Mediterranean forests", represented by mixed high coppice *Quercus ilex*- and *Quercus cerris*-dominated forests. Sites 01, 02 and 03 include nine research stations (plots) belonging to the Italian CONECOFOR programme (see Table 1). The Italian programme for the monitoring of forest ecosystems (CONECOFOR) started in 1995 within the framework of Regulation (EC) no. 1091/94 and under the UN-ECE Convention on Long-range Trans-boundary Air Pollution of the United Nations (CLRTAP).[65] A typical CONECOFOR permanent plot (Permanent Monitoring Plot, PMP) is made up of two closed

Table 1   LTER-Italy sites and their fundamental selection criteria. Selection criteria were chosen according to ILTER strategy and aims

LTER-ITALY SITES		*On-going ecological research activities*	*Long-term data series available*	*Certain funds for at least 5–10 years*	*Global strategic importance of research activities*
01	Forests of the Alps	x	x	x	x
02	Forests of the Apennines	x	x	x	x
03	Mediterranean forests	x	x	x	x
06	Southern Alpine lakes	x	x	x	
23	Lake ecosystem of Sardinia		x		x
08	North Adriatic Sea	x	x	x	x
10	Gulf of Naples	x		x	
24	Marine ecosystem of Sardinia	x			x
11	Antarctica Research Stations			x	x
17	Himalayan lakes	x	x		x

2500 m^2 areas: one was established to carry out the surveys while the other one was chosen as a control area; moreover, the outline of a typical analysis area includes a buffer zone, service facilities, a fence, meteorological stations and other sensors.

Several monitoring activities are carried out in Italian "Forest" sites: vegetation surveys, crown condition, soil chemistry, foliar chemistry, tree growth, atmospheric deposition, ozone, macro- and micro-climate, phenological observations, gas exchange, population dynamics, structure, silviculture, net primary productivity and nitrogen cycling. All CONECOFOR PMPs were initially selected not according to a systematic grid applied on Italian territory but on a preferential basis; that is, actively choosing specific areas among the most important forest communities in the country which were representative, on a wide scale, of all types of forest ecosystems for the particular climate gradient along Italy. As a consequence, it is not possible to infer general statements about forest ecosystem status at national level based on the results obtained from monitoring activities within PMPs; that is, each PMP is to be considered individually as a "case study" and the data obtained inside one plot cannot be related to the others in a statistical way.[65] Among the main priorities recognised by the integrated and combined strategy is the evaluation and quantification of: (a) the actual and potential risk status of ecosystems in the PMPs (especially in relation to atmospheric pollution and meteorological stress) and (b) the on-going change in the physical-chemical and biological system and its determinants.[65]

Given the above framework for the interpretation of scientific outputs from CONECOFOR plots, it is easy to imagine a move from monitoring issues to the search for answers to research questions and a shift from a physical-chemical approach to an ecological one, as a first step toward the investigation of status and changes in forest biodiversity.

3.3.2 Biodiversity: Ethics and Science. The Italian Academy of Forestry Science[66] identified in nature, and consequently in biological diversity, an economic value (the so-called "instrumental value") but also an "existence value" ("intrinsic value") that is the value of forest in itself, not related to human needs. Science can serve both views by the analysis of biodiversity components to understand and sustain ecological processes and services (water quality, agriculture, wood and fruits production, air quality, climate balance, *etc.*), and the "curiosity" and need for knowledge of biological actors that is "pure" research.

3.3.3 An Example: the Italian Biodiversity Research – Invertebrates. In 2003, new additional investigations related to forest biodiversity started in some CONECOFOR PMPs:[65] (a) epiphytic lichens, (b) deadwood, (c) invertebrates, (d) naturalness. In forest ecosystems, the major amount of biodiversity as regards species numbers is represented by invertebrates, especially bacteria, protozoa, molluscs, fungi, lichens and arthropods.[67] Insects are of great concern and interest at the European level: due to mono-functional silviculture practices and managing procedures, aimed at obtaining high-value wood products in the shortest time, saproxylic populations have been reduced: *Cerambyx cerdo* is considered a damaging species for woods in most forestry programmes but it is, at the same time, included in the EU Habitats Directive as end-consumer in mature forests. Twelve Italian CONECOFOR plots have been involved in a pilot project called ForestBIOTA, including a focus on invertebrates and with the principal aim of testing an effective and inexpensive protocol for insects sampling. Eight out of twelve plots included in the project are part of forest LTER-Italy facilities.

Preliminary results of the pilot project in four plots have been presented to date.[65] Samplings were conducted from June 2003 and June 2004, using a Malaise trap, a window trap and four pitfall traps in each plot (see Figure 2). Most of the collected material is still undergoing analysis, but long lists of eleven insect families have been produced so far: 21 species of *Stratiomidae*, 114 of *Syrphidae*, 206 of *Tachinidae*, 3 of *Lucanidae*, 7 of *Scolytidae* and 18 of *Rhopalocera*. Seven species of Diptera have been newly recorded for Italy (together with 28 new to Sicily and 40 new to Sardinia). Moreover, three species of *Diptera Tachinidae* were recorded as new to science (*Pales abdita, Pales*

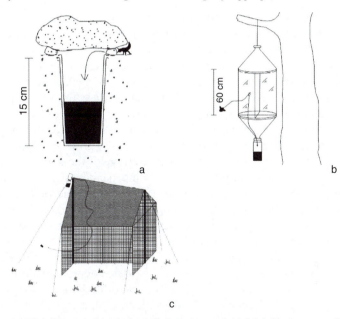

Figure 2 (a) Pitfall traps, (b) window flight traps and (c) Malaise traps used to lay out the sampling protocol for the ForestBIOTA pilot project. Many methods can be used to collect invertebrates in monitoring schemes, according to the types of information required. These kinds of trap were chosen, instead of direct sampling methods, because they fulfil several requisites: good sampling capacity, easy handling (even by non-expert people), low costs, wide use in Europe supported by literature, efficiency in collecting target *taxa*. One Malaise trap, one window flight trap and four pitfall traps were placed in each of the twelve plots surveyed. Pitfall traps were placed randomly, while Malaise and window flight traps were placed wherever suitable conditions were found. (Drawings by D. Birtele.)

marae and *Pseudogonia metallaria*), deserving special mention not only because of their discovery as newly described species but also because they seem to be strictly Mediterranean species.

3.3.4   An Example: the Italian Biodiversity Research – Vegetation.   The relationship between the number of vascular species, as a component of site biodiversity, and the characteristics of the station was investigated, relying on traditional surveys conducted in 19 PMPs (vascular plant species richness, stand structure, soil and atmosphere composition), ranging from one-storied beech and irregular-stratified Norway spruce high forests to layered oak coppice, over the period 1999–2003,[65] in order to highlight what biotic and abiotic factors influence vascular plant species richness. In fact, mean vascular species richness (including woody species) was found to be a good indicator of the total number of plant species in the plot: so, in this framework, the mean number of species per 100 m^2 was used as a response variable, *i.e.* a dependent one, while stand, soil, meteorological and deposition data were considered as the independent variables of the research. Statistical metrics were used in order to obtain a sound interpretation, because of the large number of independent variables compared to the relatively small number of plots investigated. Regarding the response of vascular plant species to chemical composition of the soil, direct as well as indirect relationships were found. In particular: (a) species richness is relatively high where there is higher content of K in the soil, high litter-fall and a large number of woody species in the dominant layer; (b) the dependent variable seems relatively low in the presence of high levels of C, N, C/N, C/P and P in the soil and high coverage of the main tree species in the upper layer. When the largest numbers of independent variables were taken into account, different results were recorded for (a) all the plots and (b) for beech plots only. When all plots were considered, the only significant correlation was with the number of tree species in the plots (that is, the higher the number of tree species in the plot, the higher the mean number of vascular plant species); when only beech plots were studied, vascular species diversity correlated positively with longitude (and, to a lesser extent, with latitude), top height, standing volume and soil N, but, on the other hand, negative correlations were found with N deposition and exceedances of a critical level of N. The extent of vascular plant diversity with regards to N content was found to be particularly interesting: in fact, especially in southern Italy, the number of vascular species increased with soil N content, but, on the other side, the number of species decreased with the increase of N deposition and exceedances of an N critical level in most cases in plots located in beech forests in northern Italy, characterised by low content of N in the soil (see Figure 3). In order to extend this analysis, the continuation of long-term data series seems to be particularly useful.

3.4 Biodiversity Status and Change in Freshwater Ecosystems: Examples from North-Italy

The international limnological literature demonstrates that there has been little research of diversity in freshwater lakes over long periods. A uniform, long-term trend in biodiversity in the world's lakes has been related to trophic change, and the corresponding effect on ecosystem biodiversity is now widely acknowledged,[68,69] accepting that there can be different responses of species according to the compartment of the trophic web and the particular ecosystem. The general trend is represented by a bell-shaped curve, with high biodiversity values corresponding to an oligomesotrophic state. Here perhaps the optimum equilibrium between the creation of new niches is reached, brought about by the increase in range and gradients of abiotic parameters and the negative effects of an excess of nutrients, such as the critical conditions for zooplankton growth and reproduction, the reduced euphotic depth or the self-shading of phytoplankton. Data for certain Italian lakes have been collected since the nineteenth century, but are generally too sparse and irregularly collected for

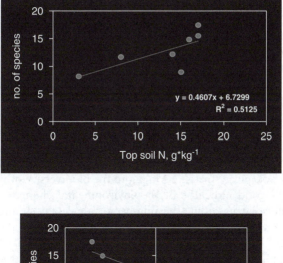

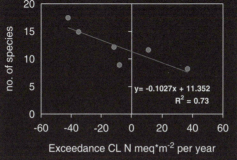

Figure 3 Vascular plant species richness is correlated (a) with nitrogen content of soil and (b) with exceedance of critical load for nitrogen, in seven CONECOFOR permanent monitoring plots in Italy, dominated by *Fagus sylvatica* forest (original, based on data from ref 85).

comparison over time. Moreover, sampling with nets was common until the 1950s and underestimated the number of species and individuals of phytoplankton. Lack of standardisation in counting methods by different investigators made it difficult to assess general changes in biodiversity in most Italian lakes (discrete samples at different depths or integrated samples over variable water column layers). Regular monitoring and standard limnological methods were established to study "cultural" eutrophication of the lakes in the subalpine district in the last 30–35 years. For many lakes, there was more emphasis on the analysis of the water chemistry than of the biota illustrated by only three papers, 1950–1980, providing complete and comparable biological data for Lake Garda, Como and Iseo. This at least allowed an analysis of long-term changes of phyto- and zooplankton. The alternation of many investigators in counting the samples and the adoption of different counting techniques further complicated the picture. Therefore, studies dealing with long-term plankton biodiversity in Italian lakes are virtually non-existent, even though the sparse existing data testify to important changes in the species structure of plankton assemblages; lakes Candia, Maggiore and Orta represent three rare exceptions in this respect. All three lakes were regularly monitored, carrying out monthly (or, sometimes, fortnightly) sampling, continuously for *c.* 25 years. In each of these lakes the long-term evolution of biodiversity has been evaluated, in particular concerning the plankton, although in some cases other biotic compartments were considered, such as macrophytes for Lake Candia. The general pattern that emerged from these analyses points towards an increase of biodiversity after the improvement of the water quality. Since the 1960s a worsening of the water quality of Lake Maggiore was recorded, due to the increased nutrient loading. Since the end of the 1970s, thanks to recovery measures, the phosphorus availability gradually decreased. In a few years, notable changes in the food web took place. In Lake Maggiore the biodiversity of phytoplankton proved to be a good indicator of the environmental changes: following the decrease of phosphorus input, the number of algal taxa gradually increased from about 50 total taxonomic units recorded at the end of the 1980s to the almost 80 taxa found in the most recent years. The most significant increase of diversity took place among the cyano-bacteria: from two or three blue-greens usually dominant in the eighties, to seven or eight dominant species in the most recent period. A comparison of samples collected during the 1980s with those collected at the end of the 1990s shows that the biodiversity increased yearly as well as seasonally. The reliability of the above findings is confirmed by the fact that the samples were counted with the same method and by the same person for the whole time series. The calculation of Shannon–Wiener diversity and of the evenness, carried out by the late Dr Delio Ruggiu, confirmed the link between the increasing phytoplankton diversity and the improving trophic status of Lake Maggiore. An analysis of the long-term zooplankton data demonstrated the same pattern of increasing biodiversity during the gradual oligotrophication of the lake.

Summarising, the limnological research in Italy offers few examples of studies dealing with the status of biodiversity and its long-term changes. The results confirm the existence of a close relationship between biodiversity and trophic evolution, a pattern already observed in many lakes worldwide. Not far from Italy, the Swiss sub-alpine lakes offer many examples of increasing biodiversity during the oligotrophication process (L. Lucerne, L. Sempach, L. Greifensee). Their trophic evolution mirrors the processes observed in the deep Italian sub-alpine lakes: for instance, very similar changes in the phytoplankton species structure and biodiversity were recorded in both Lake Maggiore and Lake Lucerne. The comparison among Swiss and Italian deep lakes seems to indicate the existence of common patterns of ecosystem reaction to the long-term trophic changes when the lakes share similar basic features (morphometry, hydrology, climate, morphology of the catchment). In general terms, the Intermediate Disturbance Hypothesis[70] seems to be appropriate also to describe the relationship between a lake's biodiversity and its trophic evolution: moving from a high degree of disturbance (eutrophication) to a lower degree (oligotrophication), the biodiversity increases. This is what, in fact, is happening in many lakes around the world. After a certain degree of oligotrophication, diversity should decrease again. However, none of those lakes seems to have reached a stable oligotrophic status yet, so we do not really know what we should expect, in terms of biodiversity, at the opposite end of the curve.

3.5 Biodiversity Status and Change in the Marine Ecosystem: Examples from the Pelagic Ecosystem in Italy

The marine environment is subject to intense human pressure and considerable changes are occurring as a consequence of over-exploitation, habitat destruction, introduction of exotic species, pollution, eutrophication and global climate change. Notwithstanding the acknowledged global importance of marine systems, the estimate of the magnitude of biodiversity at most scales is surrounded by a large degree of uncertainty. Species in the sea are far less known than on land. The major gap in the knowledge of marine biodiversity concerns those groups (prokaryotic and microbic taxa) that actually dominate the biogeochemical processes. Our perception of marine biodiversity is biased by the geographical distribution and focus of the research on few phyla: most published aquatic biodiversity research comes from few countries, focuses on relatively few specific regions and refers mainly to commercial (fish) or charismatic species (birds and marine mammals), while other organisms (*e.g.* copepods, nematodes, microbes) that play ecologically important roles are rarely mentioned.

Most research evidence and ecological theories on the effects of biodiversity on ecosystem functioning (BEF) come from terrestrial plant systems, particularly grassland. This issue has received relatively much less consideration in aquatic ecosystems. Actually, marine scientists have been very productive in research that is highly relevant to BEF issues in recent years, even though they have not placed that research within the appropriate conceptual context.

A network of excellence on Marine Biodiversity and Ecosystem Functioning (MarBEF: www.marbef.org) was established in 2004 (Sixth Framework Programme for Research and Technological Development of the EU). It comprises over 700 marine scientists from 24 countries throughout Europe and the principal aim is to support research on the relationships between biodiversity and functioning of ecosystems. The integration, in a single and unifying picture, of the mutual interactions among biodiversity changes, ecosystem functioning and abiotic factors represents the major challenge in biodiversity research. The (re-)analysis of large-scale and long-term studies, even though they were not framed in the BEF context, represents a valuable way to explore and evaluate biodiversity and ecosystem changes. In this context, ecological time-series are fundamental tools to trace the long-term variations of marine ecosystems, detect significant shifts and assess whether changes are attributable to human or natural causes. Moreover, long-term series are essential for testing ecological theories, for enhancing our limited capacity for short- and medium-term forecasting and for managing the resources.

3.6   Toward a Core Set of Biodiversity Indicators for LTER Sites

It is highly desirable to develop a core set of biodiversity indicators for LTER sites which may be applied on the basis of harmonised methods at pan-European level, across different bio-geographical regions. The Pan-European Biological and Landscape Diversity Strategy (PEBLDS) was developed to support the implementation of the UN Convention on Biological Diversity at the pan-European level, by the initiatives of the Council of Europe and the United Nations Environment Programme (UNEP). In this framework, the biodiversity resolution taken by the 5th Conference of the European Ministers of Environment "Environment for Europe" (Kiev, 2003) includes the keystone decision to develop by 2006 a core set of biodiversity indicators and to establish by 2008 a pan-European network on biodiversity monitoring and reporting, with a framework of collaboration with MCPFE (Ministerial Conference on the Protection of Forests in Europe). A pan-European Co-ordination Team, formed by the European Environment Agency, UNEP, the European Centre for Nature Conservation and the Expert Groups leaders has operated since 2004, having initiated its work collecting available information. The work plan elaborated, approved and funded by the EC,[71] provides the logical framework for the activities that need to be carried out in order to ensure European coordination of the development and implementation of biodiversity indicators. The indicators will be applied in assessing, reporting on and communicating achievement of the 2010 target to halt biodiversity loss. This activity is called "Streamlining European 2010 Biodiversity Indicators (SEBI2010)".[72] The Expert Groups have started their activities; demonstration activities to be carried out in three test countries are expected in 2006; by 2007 the definition and publishing of the final revision of the indicator set; and by 2008 the establishment of a co-operative monitoring network at pan-European level. The selected "headline" indicators are based on the set currently proposed at global and European level and just included in an official EU list (Annex 2[71]): trends in abundance and distribution of selected species, change in status of threatened and/or protected species, trends in extent of selected biomes, ecosystems and habitats, trends in genetic diversity of domesticated animals, cultivated plants and fish species of major socio-economic importance, coverage of protected areas, nitrogen deposition, number and costs of invasive alien species, impact of climate change on biodiversity, marine trophic index, connectivity/fragmentation of ecosystems and water quality in aquatic ecosystems.

A good example of results and work in progress under the SEBI2010 initiative is the overall headline indicator called "Trend in extent and composition of selected ecosystems", currently under development by the SEBI2010 Expert Group 2. A specific "Forest Area Indicator" is ready for implementation, mainly based on quantitative data (trend of forest area, considering forest types), but for a proper understanding and evaluation it needs to be complemented by a qualitative indicator, taking into account status and trends of key characteristics of forest ecosystems, a "Forest Status Indicator (FSI)". Development of FSI is based on detailed collection of available meta-data and harmonised methods (EU Forest Focus and UN-ECE ICPs, National Forest Inventories, Natura2000 National Reports, MCPFE Reports, *etc.*). It will consist of a synthesis from surrogate measures (sub-indicators) for biodiversity (tree condition, deadwood amount and type, plant species composition, *etc.*) per forest type in Europe, with the aim to evaluate results provided by the Forest Area Indicator, taking into account concepts like quality, functionality and integrity of forest ecosystems. It will be based on sub-indicators identified at pan-European level (4th Ministerial Conference on the Protection of Forests in Europe, MCPFE) and implemented at pan-European (EU Forest Focus and UN-ECE ICP Forests) and National level (NFIs) as follows: (a) EU Forest Focus and UN-ECE ICP Forests Level I: tree condition data on *c.* 3000 points, since 1985 (continuously for 20 years); forest structure, deadwood and plant species composition on *c.* 6000 points, since 2007 (pilot project BioSoil); (b) EU Forest Focus and UN-ECE ICP Forests Level II: tree condition data on *c.* 700 plots and plant species composition on *c.* 500 plots, since 1995 (continuously for 10 years); deadwood data on *c.* 100 plots, since 2006 (pilot project ForestBIOTA); (c) National Focal Points: tree species composition and dead-wood data from a number of NFIs all over Europe; (d) Natura2000 National Reports: "conservation status" of a number of SCIs (47% of them including forests) all over Europe; (e) MCPFE Reports and National data: "protected forests" amount. Data will be organised according to a revised and improved version of FTBA (BEAR Forest Types for Biodiversity Assessment), recently released to EEA. FSI development meets the requirements of SEBI2010 concerning delivering data on changes in the time of some key attributes of forest ecosystems in Europe; the emphasis on the qualitative aspects of biodiversity is policy relevant for the management of the environment. Most of the data are harmonised at pan-European level; in some cases they cover a period of 20 years, according to a systematic network well representative of all Europe and are readily available from international bodies (EU and UN-ECE). There is the possibility for up- and down-scaling of data collected at Level I and Level II sites. FSI will be based on sub-indicators broadly accepted; it is very sensitive, being able to detect changes in time frames and on scales that are relevant to the decisions. It can be updated regularly, if adopted at European level, on the basis of routine monitoring programmes. The available data are consistent in space and cover most of the EEA countries. The indicator could be represented by star diagrams, including all sub-indicators/forest types/years (each diagram per each available year). Changes in the time and "distance" from target values can be easily recognised by the change in shape of the diagrams. Some examples for beech and spruce forests in Italy are given in Figure 4.

4   Indicators of Biodiversity and its Change for the "Wider Countryside"

A convincing justification for the construction of an integrated and harmonised network of LTER sites across Europe has been presented. However, this should

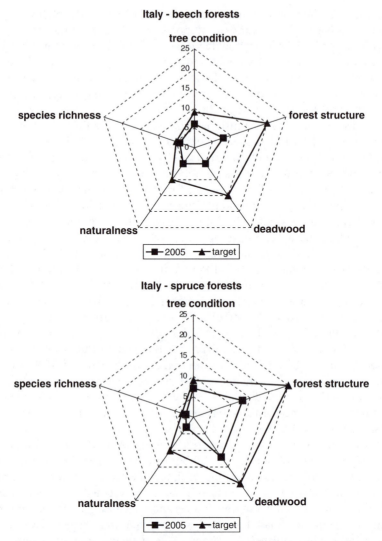

Figure 4 Examples of star diagrams based on same sub-indicators of *Forest Status Indicator* (original data for 2005, partially from refs 85 and 86).

not exclude the contribution of further sources of biodiversity assessment over time, particularly those that have been established for longer periods.

4.1 Limitations of Discrete, Intensively Monitored Locations

Realistically, a future integrated and harmonised pan-European LTER network will achieve only relatively few locations, and these will necessarily be small in area. The capital and maintenance costs of infrastructure and instrumentation and the labour costs for the collection of measurements will be high, and hence will restrict the overall coverage of habitats and regions. These detailed measurements will be restricted to a relatively discrete list of habitats and species, selected for pan-European comparison (*e.g.* headline biodiversity indicators; EEA 2004[73]). The target habitats and species will by necessity be wide-ranging rather than restricted in range or extent; neither rare nor locally abundant, respectively. Due to the demand for regular access, LTER sites may also be selected according to factors such as ease of access or willingness of land owners to agree access, culminating in a biased rather than entirely scientifically objective stratified sampling design. A disadvantage of fixed sites of relatively small area is the effects of disturbance from the regular, continuous (sometimes destructive) sampling of various plant and animal populations and physico-chemical parameters on target organisms of interest. The lack of suitable taxonomic expertise to undertake the identification of particular plant and animal taxa severely limits the potential range of organisms that can be included in LTER monitoring.

4.2 Advantages of Supplementary Monitoring in the Wider Landscape

The measurement of ecological indicators is used either to assess the condition of the environment or to reveal the cause of environmental change.[74] The precise composition of such indicators is dealt with earlier, but monitoring for ecosystem or resource management often requires data collected on such indicators at specific sites. In contrast, policy decision-makers require infor-mation across broader geographical areas and the site data may not be representative of the broader state of the environment.[74] The network of proposed LTER sites can only ever cover a very minor fraction of the entire European land surface. Whilst the "wider countryside" or "landscape matrix" is clearly enormously important for supporting most biodiversity, data col-lected on a selected taxonomic group and collected from a broad-ranging but truly random and stratified sample may be more representative of general habitat condition than LTER sites. European countries have varied numbers of pre-existing long-term environmental or ecological datasets from numerous locations in the wider countryside.[75] These often represent longer-term data and illustrate trends for a different set of organisms compared with those of the relatively restricted list under surveillance at LTER sites. These sources often contribute long-term data for rare or locally abundant species within different regions, according to nature conservation status, the availability of taxonomic expertise or special interest groups. These data allow the possibility to test the broader consistency of trends recorded for one or more of the habitats and species under surveillance at LTER sites in the broader context, which is of relevance to policy makers. There is also the potential to demonstrate such consistency of trends for a different set of species and habitats across land-scapes under "real" management and a range of environmental contexts.

Examples could be the monitoring of the condition of sites under the EC Habitats Directive protected sites, namely Special Areas for Conservation (SAC), which offers the possibility to include the extent and condition of key habitats and certain associated species of conservation concern. National monitoring of aquatic pollutants and invertebrates is carried out by European national environmental protection agencies in accordance with the EC Waters Directive (*e.g.* Harmonised Monitoring Scheme). Natural history records of special interest groups also generate inexpensive but adequate data from across the broader countryside:

- Voluntary effort recording a broad range of habitats and species, coordinated by Local Government Record Centres.
- National surveys of major charismatic wildlife groups by volunteers with professional guidance, coordinated by NGOs, *e.g.* BirdLife International partners in several European countries and Butterfly Conservation in the UK.

4.3   Data Sources that could Contribute to Surveillance in the Broader Landscape

The existence of inherited data, ease of access, comparability, both across countries and with data collected from the LTER, and cost-effectiveness all are major considerations for monitoring in the broader countryside. Remote sensing offers potentially full coverage for the assessment of the pattern and extent of recognisable land cover classes/broad habitats.[75] It may be possible to lobby for greater general vegetation survey through appropriate training of farmers to undertake self-assessment of land in receipt of agri-environment incentive payments, possibly the presence or absence of target species in agricultural land. The monitoring of charismatic/special interest groups encouraged and coordinated by NGOs (*e.g.* birds, butterflies, moths, dragonflies and flowering plants) has been a major contributor to time-series data and data of broad geographical coverage, although only in a few European countries.[75] Common Standards Monitoring (CSM) is used by the conservation agencies to assess whether designated sites, such as EC Habitats Directive SACs, are in "favourable condition". There has been an increasing emphasis on monitoring the condition of designated conservation sites, including a need to underpin the relatively superficial assessments made under CSM with more detailed, scientific measurements at a sub-set of sites. English Nature conducted a pilot study for a "validation network" to provide this function.[76] Monitoring encouraged by Local Government, such as the National Biodiversity Network in the UK (http://www.nbn.org.uk), has only been established in recent years. However, they have also greatly improved accessibility to existing data, especially for habitats and species on lists that have been agreed for the National biodiversity strategy. The ICP Forests Level II Programme monitors a wide range of variables relating to air pollution and climate and their impacts on production forests, although lacks any measurement of forest habitats and species. The assessment of waterways is essential because water courses integrate pollution effects in the broader landscape, *e.g.* land management effects, atmospheric deposition of pollutants and the potential buffering effect of soils and geology of particular river catchments. The Acid Waters Monitoring Network includes both physical and biological variables to investigate the effects of acidifying atmospheric deposition on freshwater systems and their catchments.[77]

The main challenges in using data derived from the diverse sources of measurement in the broader countryside are:

- consistency in the taxonomic groups measured
- use of comparable methods
- substantial duration and overlap of time-series data
- similar frequency of recording
- availability of associated physico-chemical measurements.

The nature of the source data, collected for a particular local or national purpose, and the need for consistency in the method of measurement for a particular time-series, makes it difficult to consider any changes. Such diverse monitoring will not attain the same degree of standardisation that should be the major characteristic of monitoring under a harmonised and integrated set of measurements carried out at a European network of LTER sites. The collection of additional biodiversity data in the wider countryside may also have the additional benefit of greater relevance and proximity to more of the European public than the selected LTER sites. Hence, it may facilitate greater public understanding and appreciation of pressures on biodiversity and their consequences. This important area is explored in the next section.

5   How to Communicate Biodiversity Assessments to Stakeholders and the Public?

The results of scientific research rarely become visible to the general public or even to decision makers. This has to be changed, as good science is not enough to halt biodiversity loss; the participation of society is indispensable.[78] Awareness and appreciation of biological diversity is highly dependent on the extent to which science on this matter is perceived and accepted by the public and different user groups. Effective science communication can change society's attitude to nature itself and the services it provides.[79]

The communication of biodiversity assessment is different from other types of science communication in several respects. The aim is not just to inform people on scientific achievements, but to motivate a change in their attitudes and behaviour, such as consumption patterns. To do so, people need reasoning and convincing information, and a learning process.[79] Knowledge, awareness and attitudes are not enough to effect behaviour change. Emotions and moral convictions also influence actions[80] and have to be taken into consideration in the communication process. The loss of biodiversity is rarely appreciated by the public; a lack of information and data on the change is a major element. The necessary data must be derived from existing and future biodiversity assessments and monitoring programmes. Generally, the efficiency of communication of the results from these activities to users and the public needs strengthening.

The importance of science communication and awareness raising is acknowledged by several international efforts. The Convention on Biological Diversity expressed this (Article 13. Public education and awareness) at a global level. Lately the European Commission has identified the building of public education, awareness and participation for biodiversity as a supporting measure to halt the loss of biodiversity.[71] Campaigns and communication programmes have to be launched and access to information has to be ensured to fulfil the requirements. Effective communication methods can support this process. This section summarises general aspects of communication and specific methods for biodiversity assessments.

5.1   General Principles

The popularisation of science should be treated as seriously as science itself. This is the basis for success, but it is generally not happening. The reward system of scientific research, where the number and impact of papers in scientific journals are considered, does not support this approach. So the first step is to put more emphasis on science communication and popularisation in the career-building of scientists. This would automatically result in an increase in respect for the audience that could help to establish a basis for trust.

Three main communication strategies exist: values-based communication, strategic-frame analysis and social marketing.[80] Values-based communications analyse personal beliefs and valuation of qualities of the target audience and try to persuade by reason and motivate through emotion. Strategic-frame analysis uses cognitive sciences to identify how messages are encoded to be interpreted in relation to existing beliefs. Social marketing considers the desired activity of a product that has a price and needs promotion. All three approaches need data and information in the process to support the message. This is where biodiversity assessment results can be used in biodiversity conservation issues.

5.2   Setting Objectives

The overall aim is not just to inform people, but to motivate change in their attitudes and behaviour, which is the most difficult activity. The task is to use the main findings of biodiversity assessment and monitoring programmes as part of the communication campaign. The results of these often have implications for policy issues, so the relation between biodiversity matters and decisions or legislation represents further information that must be communicated. In most cases the objective is to raise attention and awareness by understanding the concept and the importance of biodiversity. The ultimate target of an increasing number of campaigns is to motivate action, change behaviour and involve people in problem-solving in order to halt biodiversity loss. Biodiversity is a complex issue, hardly understood unless elements, levels and defined aspects are communicated. A focus can be put on landscape, community, species, population or genetic levels.

When setting the objectives of communication, the message to be transmitted has to be clearly defined. General messages can be: biodiversity is life; biodiversity is declining; biodiversity is vulnerable; biodiversity loss has direct implications for human life; we are all responsible; everyone can do something about it. Data and information to support the message have to be selected and used in the campaign. Message development should be based on what the intended audience values. Key messages should be positive, simple to understand, memorable, accurate and realistic.[81]

5.3   Selecting the Target Audience

To influence biodiversity loss, focus is required on communication with key actors, who make decisions at a large scale. Usually emphasis is on school children, a group with no direct influence on implementation processes.[79] However, their future role and influence on consumption patterns is evident. Therefore, the scope of the target audience has to be well adjusted to the aims.

The communication approach should greatly differ for different audiences as the perception of and attitude to the message largely depends on what they value. Policy makers need links to their interests in very short messages, supported by a few numbers. Other stakeholders usually need more detail, especially if the message concerns their own subject area. For other scientists the scientific approach that generated the information is of more importance. It is hard to convince the business community to support biodiversity matters, unless beneficial to the market (biodiversity "brand" used[82]). Land managers and farmers are an important component of the audience because of their direct impact on land use. Social market methodology to promote sustainable agriculture turned out to be effective to bridge the existing gap between research and managers.[80]

Teachers have an important role to play in distributing the message to the younger generation. They require special support for this activity. The rural population needs a different approach from that of the urban population or the general public. Families accept messages more if all members find interesting elements. Children need age-specific approaches. The main rule is that the communicator should know as much as possible about the views and attitudes of the target audience in advance.

5.4   Selection of Appropriate Tools

Three major types of tool exist: one-way, two-way or interactive methods. One-way communication means one-way information flow from the communicator to the audience. Such tools are printed materials, web sites, CDs, posters, exhibitions, films, comics or the media. In this case very little is known about the impact of the effort. Interactive methods are always more effective, but more labour- and time-consuming. Two-way communication can be conducted by personal encounters or by written materials. Personal meetings are more effective but generally reach fewer people. Such meetings can be open days at institutes, environmental education campaigns, field visits, training pro-grammes, practical activities, *e.g.* monitoring, volunteer work, nature camps, visitor-centre activities. Larger groups can be involved through the media in international days for biodiversity or nature. The coordination of master's theses in the field of biodiversity assessment and monitoring can also contribute to awareness-raising. The use of questionnaires, internet forums and interactive software are widely used tools for two-way communication with larger audiences; however, participation might be surprisingly low. Evaluation of effectiveness should go parallel with both types of communication tools.

5.5   How to Do It?

Two main tasks can be identified during the communication process: the first is to capture the attention of the right audience; the second is to deliver an understandable and credible message that influences the beliefs of the audience. Real-life examples help to get closer to the audience and capture attention. Difficult topics within biodiversity assessment should be related to ordinary people's lives to maintain interest. Stressing the link of natural to cultural values is a helpful option. Story-telling and dramatic examples can also be effective if appropriate for the target group. Interactive approaches or inter-esting questions put to the audience are also effective means. Get people involved directly in the process, listen to their views and suggestions. Reward positive behaviour rather than punish negative behaviour. To deliver the message, the use of easily understood indicators or surrogates of biodiversity can be helpful. Never compromise the scientific credibility, avoid too much technical jargon and use simple language. Enthusiasm, personality and cred-ibility in personal contacts are essential. Instead of details, stress principal findings and implications. Adjust the language to the target audience.[79] Do not over- or under-estimate the potential of communication; consider communica-tion barriers.[83]

5.6   Evaluation of Success

Two types of evaluation are required: the assessment of the communication process and the estimation of the impact of the message. Evaluation of the process itself helps to improve delivery in the future. Baseline values are required before starting the project; therefore the evaluation planning must be parallel to the planning of the communication activity, and the evaluation should be on-going throughout the programme.[84] Bear in mind that the best indicators of success might be the more difficult data to collect. Evaluation should be associated with the main messages and depend on the audience and the tools used.

A general method is to prepare questionnaires, not exceeding two pages, for different groups. Prizes are effective at increasing the number of respondents. Open forums and interviews can provide the opportunity for people to express their opinions if they find the topic interesting. Opinions should be assessed on what people think they can do for biodiversity before and after the programme. Organisation of competitions with the help of teachers may increase attention within the target group. Feedback can be collected through voting on the internet or phone calls from the audience (for media appearance). For large campaigns statistical data (consumption patterns/daily choices) or policy responses (new legislation) can be used as indicators of effectiveness.

5.7 Case Study: Visitor Centre in Vácrátót, Hungary

An increasing number of papers and projects deal with or give advice on biodiversity communication (*e.g.* Goldstein,[79] Farrior[80]), but no references or examples exist to our knowledge which focus on biodiversity assessment. This is usually part of the communication process providing evidence for loss of biodiversity. However, here we describe a case study that includes an assessment of biodiversity.

A visitor centre is under construction at the Institute of Ecology and Botany within the botanic garden in Vácrátót (Hungary). The aim of the centre is to provide information on the role of plants in our lives and in the functioning of global processes and to raise awareness and individual responsibility. Besides explaining biodiversity loss, the exhibition will select examples of survey, monitoring and research results from the Institute, including biodiversity assessments. The target audience is the general visitor to the garden (about 50 000 people per year). This includes school and pensioner groups, families, students, *i.e.* all ages of different social and educational levels.

The idea of biodiversity loss is symbolised by a balloon of the globe with a basket full of people and with sacks hanging down from the basket that represent pressures, like habitat loss, biological invasion, *etc.* Each sack – pressure – is detailed separately with examples and research results (see Figure 5). The National Habitat Map Database will be used to demonstrate fragmentation of habitats. Temporal changes are visible with the help of historic habitat reconstructions in the form of map series from the eighteenth century onwards. The loss of natural habitats and their patterns are similar to old codex pages that lack certain parts of the paintings and text. This approach provides the opportunity to mention restoration methodologies for habitats and codex pages as well. Surveys on invasive plants represent species level assessments. Guidance on how to use the knowledge in everyday life will be presented.

Figure 5 Conceptual sketch of the balloon installation in a biodiversity science visitor centre. The visualisation of the pressures on biodiversity as ballast sacks to the balloon illustrates the fact that these pressures have to be limited or precisely balanced to keep the balloon in flight. In the sacks, visitors will find the components of each pressure with detailed information on them.

6   Conclusions

The assessment of biodiversity status and trends is today an established component of environmental observation and monitoring programmes, performed from regional to global scales. Although the term biodiversity expresses a very complex phenomenon – the "overall richness of life on earth" – which has to be referred to at different hierarchical levels, we have demonstrated that there are adequate means and methods available to describe and quantify it. In particular, numerical indices that summarise the diversity and evenness of species distributions, environmental surrogates of biodiversity, indicator species, combined and aggregated indicators (which may comprise both species- and habitat-level information) are available and have proved useful in particular instances of the assessment of biodiversity. A substantial body of scientific literature describes the limitations, biases and pitfalls of the various measures. In practice, most assessments of biodiversity rely on the simplest and most intuitive method, consisting of species counts and abundance estimates.

Long-Term Ecological Research sites are crucially important "test laboratories" to develop and apply methods for the assessment of biodiversity. LTER sites offer the best possibility to integrate the analysis of abiotic and biotic data with the objective of identifying the relative importance of pollution or climate as drivers of change. Networking such sites across Europe and up to the global scale will reinforce and increase the certainty of such results. Our case study of the Italian LTER demonstrates the usefulness of combining continued (long-term) observation with a network of sites covering the key ecosystem types within a country or region. With this observation approach, it is possible to detect patterns which would not emerge on the basis of less intensive or temporally less-extended observation.

However, LTER sites themselves cover only a very small part of the overall landscape supporting biodiversity, and they have a bias to natural or near-natural ecosystems. It is also important to have reliable information on biodiversity status in the wider countryside. Information obtained from remote sensing offers great potential here and several processes and programmes support biodiversity assessment in the wider countryside, most importantly at the European scale the monitoring of the condition of sites under the EC Habitats Directive.

Finally, understanding and appreciation of biological diversity is highly dependent on successful science communication. We have demonstrated here that approaches must duly consider the different user groups involved. Visitor centres with a high degree of interactive exploration can be a successful means of communicating the science examining biodiversity.

References

1. UNEP, United Nations Environment Programme, "Environmental Data Report", Blackwell, Oxford, 1987.
2. WCED, "Our Common Future", Oxford University Press, Oxford, 1987.
3. I. F. Spellerberg, "Monitoring Ecological Change", Cambridge University Press, Cambridge, 1991.
4. F. B. Goldsmith (ed.), "Monitoring for Conservation and Ecology", Chapman and Hall, London, 1991.
5. D. W. Rhind, B. K. Wyatt, D. J. Briggs, J. Wiggins. "The creation of an environmental information system for the European Community", Nachrichten aus Karten und Vermessungswesen, 1986, Series 11, **44**, 147.
6. N. E. Stork and M. J. Samways, in "UNEP Global Biodiversity Assessment", V. H. Heywood (ed.), Cambridge University Press, Cambridge, 1995, 435–544.
7. K. J. Gaston and J. I. Spicer, "Biodiversity – an Introduction", Blackwell, Oxford, 1998.
8. P. Crane and P. Bateson, "Measuring Biodiversity for Conservation", The Royal Society, London, 2006, 1–56, http://www.royalsociety.org/display-pagedoc.asp?id=6712.

9. M. A. Smith, B. L. Fisher and P. D. N. Hebert, *Phil. Trans. Roy. Soc.*, 2006, **B 360**, 1825–1834.
10. K. J. Gaston, "Biodiversity: a Biology of Numbers and Difference", Blackwell, Oxford, 1996.
11. J. Puumalainen, "Forest biodiversity – assessment approaches for Europe", Ispra, European Commission Joint Research Centre, 2002.
12. C. Papazoglou and V. Phillips, "Biodiversity indicator for Europe: population trends of wild birds", 2003, http://www.rspb.org.uk/Images/ Biodiversity%20indicators.
13. M. de Heer, V. Kapos and B. ten Brink, "Biodiversity trends and threats in Europe", RIVM report 717101001, 2005, 3–72, Bilthoven, RIVM.
14. R. H. MacArthur, *Am. Nat.*, 1964, **98**, 387–397.
15. E. R. Pianka, *Ecology*, 1967, **48**, 333–350.
16. M. Game and G. F. Peterken, *Biol. Conservat.*, 1984, **29**, 157–181.
17. D. D. Murphy and B. A. Wilcox in, "Wildlife 2000" "Modelling habitat relationships of terrestrial vertebrates", in J. Verner, M. L. Morrison and C. J. Ralph (eds), Univ. Wisconsin Press, Madison, 1986, 287–292.
18. R. F. Noss, *Conservat. Biol.*, 1990, **4**, 355–364.
19. T. M. Caro, *Anim. Conservat.*, 2003, **6**, 171–181.
20. P. B. Landres, J. Verner and J. W. Thomas, *Conservat. Biol.*, 1988, **2**, 316–328.
21. T. M. Caro and G. O'Doherty, *Conservat. Biol.*, 1999, **13**, 805–814.
22. D. Simberloff, *Biol. Conservat.*, 1998, **83**, 247–257.
23. J. M. Roberge and P. Angelstam, *Conservat. Biol.*, 2004, **18**, 76–85.
24. J. M. Favreau, C. A. Drew, G. R. Hess, M. J. Rubino, F. H. Koch and K. A. Eschelbach, *Biodiversity and Conservation*, 2006, in press.
25. M. A. McGeoch, *Biol. Rev.*, 1998, **73**, 181–201.
26. P. Duelli and M. K. Obrist, *Agr. Ecosyst. Environ.*, 2003, **98**, 87–98.
27. C. R. Margules and R. L. Pressey, *Nature*, 2000, **405**, 243–253.
28. M. C. D. Speight, "Saproxylic Invertebrates and their Conservation", Council of Europe, Strasbourg, 1989, 81.
29. K. Schiegg, *Ecography*, 2000, **23**, 579–587.
30. J. Siitonen, *Ecological Bulletins*, 2001, **49**, 11–41.
31. WWF, "Deadwood – living forests", WWF-Report, 2004, 19.
32. J. Schmidl and H. Bußler, *Naturschutz und Landschaftsplanung*, 2004, **36**, 202–218.
33. K. Alexander, "The British saproxylic invertebrate fauna – People's Trust for Endangered Species", 2003, 3.
34. B. Økland, A. Bakke, S. Hågvar and T. Kvamme, *Biodiversity and Conservation*, 1996, **5**, 75–100.
35. P. Martikainen, L. Kaila and Y. Haila, *Conservat. Biol.*, 1998, **12**, 293–301.
36. B. G. Jonsson and M. Jonsell, *Biodiversity and Conservation*, 1999, **8**, 1417–1433.
37. T. Ranius, *Biodiversity and Conservation*, 2002, **11**, 931–941.
38. A. Juutinen, M. Mönkkönen and A. L. Sippola, *Conservat. Biol.*, 2006, **20**, 74–84.

39. G. Mikusinski, M. Gromadzki and P. Chylarecki, *Conservat. Biol.*, 2001, **15**, 208–217.
40. R. Bütler, P. Angelstam, P. Ekelund and R. Schlaepfer, *Biol. Conservat.*, 2004, **119**, 305–318.
41. T. Ranius and L. Fahrig, *Scandinavian Journal of Forest Research*, 2006, **21**, 201–208.
42. A. Shmida and M. V. Wilson, *J. Biogeogr.*, 1985, **12**, 1–21.
43. C. Rahbek, *Ecol. Lett.*, 2005, **8**, 224–239.
44. E. Fleishman, C. J. Betrus and R. B. Blair, *Landscape Ecology*, 2003, **18**, 675–685.
45. G. R. Hess, R. A. Bartel, A. K. Leidner, K. M. Rosenfeld, M. J. Rubino, S. B. Snider and T. H. Ricketts, *Biol. Conservat.*, 2006, **132**, 448–457.
46. E. J. Pharo and A. J. Beattie, *Aust. J. Ecol.*, 1997, **22**, 151–162.
47. T. H. Ricketts, E. Dinerstein, D. M. Olson and C. Loucks, *Bioscience*, 1999, **49**, 369–381.
48. B. Reyers and A. S. van Jaarsveld, *South African Journal of Science*, 2000, **96**, 406–408.
49. D. P. Faith, *Ecography*, 2003, **26**, 374–379.
50. S. Sarkar, J. Justus, T. Fuller, C. Kelley, J. Garson and M. Mayfield, *Conservat. Biol.*, 2005, **19**, 815–825.
51. V. Kati, P. Devillers, M. Dufrene, A. Legakis, D. Vokou and P. Lebrun, *Conservat. Biol.*, 2004, **18**, 667–675.
52. M. Sætersdal, I. Gjerde, H. H. Blom, P. G. Ihlen, E. W. Myrseth, R. Pommeresche, J. Skartveit, T. Solhøy and O. Aas, *Biol. Conservat.*, 2004, **115**, 21–31.
53. N. Sauberer, K. P. Zulka, M. Abensperg-Traun, H.-M. Berg, G. Bieringer, N. Milasowszky, D. Moser, C. Plutzar, M. Pollheimer, C. Storch, R. Tröstl, H. G. Zechmeister and G. Grabherr, *Biol. Conservat.*, 2004, **117**, 181–190.
54. A. Balmford, P. Crane, A. Dobson, R. E. Green and G. M. Mace, *Phil. Trans. Roy. Soc. London B*, 2005, **360**, 221–228.
55. M. de Heer and V. Kapos, and B. J. E. ten Brink, *Phil. Trans. Roy. Soc. London B*, 2005, **360**, 297–308.
56. R. D. Gregory, A. van Strien, P. Vorisek, A. W. G. Meyling, D. G. Noble, R. P. B. Foppen and D. W. Gibbons, *Phil. Trans. Roy. Soc. London B*, 2005, **360**, 269–288.
57. S. H. M. Butchart, A. J. Stattersfield, L. A. Bennun, S. M. Shutes, H. R. Akcakaya, J. E. M. Baillie, S. N. Stuart, C. Hilton-Taylor and G. M. Mace, *PLoS Biology*, 2004, **2**, 2294–2304.
58. H. Kaufmann and M. Anderson, "ILTER Strategic Plan", Environment and Enterprise Strategies LLC, 2006.
59. A. Vadineanu, S. Datcu, M. Adamescu and C. Cazacu, "The state of the art for LTER activities in Europe", ALTER-Net WP I3 Report, University of Bucharest, Department of Systems Ecology, 2005.
60. ALTER-Net web site, www.alter-net.info.
61. W. De Vries, M. Forsius, M. Lorenz, L. Lundin, T. Haussmann, S. Augustin, M. Ferretti, S. Kleemola and E. Vel, "Cause-effect relationships

of forest ecosystems", Joint report by ICP Forests and ICP Integrated Monitoring, UN-ECE, BFH, SIKE, 2002, 46.

62. UN-ECE ICP Forests web page, http://www.icp-forests.org.
63. UN-ECE ICP IM web page, http://www.vyh.fi/eng/intcoop/projects/icp_im/im.htm.
64. R. Fischer (ed.), "The Condition of Forests in Europe", 2006 Executive Report, UN-ECE, Geneva, 2006, 33.
65. M. Ferretti, B. Petriccione, G. Fabbio and F. Bussotti (eds), "Aspects of biodiversity in selected forest ecosystems in Italy, status and changes over the period 1996–2003", Third report of the Task Force on Integrated and Combined (IandC) evaluation of the CONECOFOR programme, Annali Istituto Sperimentale per la Selvicoltura, Special Issue (Arezzo), 2006, **30**, Suppl. 2, 107–111.
66. A. Barbati, P. Corona and M. Marchetti (eds), "Definizione delle linee guida per la gestione ecosostenibile delle risorse agrosilvopastorali nel Parchi nazionali", Completamento del quadro metodologico e conoscitivo, Ministero dell'Ambiente, Servizio Conservazione della Natura, Accademia Italiana di Scienze Forestali, Firenze, 2001, 88.
67. O. Ciancio and S. Nocentini, "La conservazione della biodiversità nei sistemi forestali, 1. Ipotesi per il mantenimento degli ecosistemi", L'Italia Forestale e Montana, Rivista di Politica Economia e Tecnica, Anno LVII, Numero 6, Novembre–Dicembre, 2002.
68. C. S. Reynolds, "Vegetation processes in the pelagic, a model for ecosystem theory", Ecology Institute, Oldendorf/Luhe, Germany, 1997.
69. R. G. Wetzel, *Arch. Hydrobiol. Spec. Issues Advanc. Limnol.*, 1999, **54**, 19.
70. J. Connell, *Science*, 1978, **199**, 1304.
71. European Commission, "Halting the loss of biodiversity by 2010, and beyond. Sustaining ecosystem services for human well-being", Communication from the Commission no. COM(2006) 216, Bruxelles, 2006, 15, with 2 annexes.
72. European Community Biodiversity Clearing House Mechanism, SEBI-2010, 2006, http,//biodiversity-chm.eea.europa.eu/information/indicator/F1090245995.
73. European Environment Agency, EU Headline Biodiversity Indicators, Proceedings of the Stakeholders' Conference, "Biodiversity and the EU – Sustaining Life, Sustaining Livelihoods", Grand Hotel, Malahide, Ireland, 25–27 May 2004.
74. G. J. Niemi and M. E. McDonald, *Annu. Rev. Ecol. Evol. Systemat.*, 2004, **35**, 89–111.
75. M. Bredemeier, "Biodiversity Assessment and Change. Activities and Methods of Biodiversity Assessment and Monitoring in European Countries", Sixth Framework Programme, Global Change and Ecosystems, Network of Excellence, A Long-term Biodiversity, Ecosystem and Awareness Research Network (ALTER-Net) publication WPR2-2005-01, FERC, Forest Ecosystems Research Center of Goettingen University, Germany, 2005.

76. C. E. Bealey and J. Cox, "Validation network project: upland habitats covering blanket bog, dry dwarf shrub heath, wet dwarf shrub heath, Ulex gallii, dwarf shrub heath", English Nature Research Report, 2004, 564, English Nature, Peterborough.

77. M. D. Morecroft, A. R. J. Sier, D. A. Elston, I. M. Nevison, J. R. Hall, S. C. Rennie, T. W. Parr and H. Q. P. Crick, "Targeted Monitoring of Air Pollution and Climate Change Impacts on Biodiversity", Final report to Department for Environment, Food and Rural Affairs, Countryside Council for Wales and English Nature (CR0322), 2006.

78. J. Lawton, "European biodiversity conservation, science, policy and practice", 1st European Congress on Conservation Biology, Book of Abstracts, 2006, 7.

79. W. Goldstein, "Some lessons on communicating biodiversity and follow up actions", IUCN, 2005, http//www.iucn.org/themes/cec/biodiversity/Com_ed_for_bd_workshopJan2005.pdf.

80. M. Farrior, "Breakthrough Strategies for Engaging the Public, Emerging Trends in Communications and Social Science", Biodiversity Project, 2005, http//www.biodiversityproject.org/bpemergingtrendspaper.doc.

81. LIFE Focus, "LIFE Focus/LIFE-Nature, communicating with stakeholders and the general public – Best practice examples for Natura2000", European Commission, 2004, http//europa.eu.int/comm/environment/life/infoproducts/naturecommunicating_lowres_en.pdf.

82. T. Kitchin, "Assuring biodiversity. A brand-building approach", The Glasshouse Partnership, London, 2004, URL, http//www.glasshousepartnership.com/branding.

83. IUCN Commission on Education and Communication, 2006, URL, http//www.iucn.org/themes/cec/principles/dos.htm (06. 09. 2006).

84. J. Metcalfe and D. Perry, "The evaluation of science-based organisations' communication programs", Presentation to Australian Science Communicators conference, Sydney, 2001, http//www.econnect.com.au/news_papers.htm#eval.

85. M. Ferretti, B. Petriccione, G. Fabbio and F. Bussotti (eds), "Aspects of biodiversity in selected forests ecosystems in Italy, status and changes over the period 1996–2003", Third Report of the Task Force on Integrated and Combined (IandC) evaluation of the CONECOFOR programme, Annali Istituto Sperimentale per la Selvicoltura, Special Issue, 2006, Vol. 30, Supplemento 2, 2006, 112.

86. B. Petriccione, in "Monitoring and Indicators of Forest Biodiversity in Europe. From Ideas to Operationality", M. Marchetti (ed.), EFI Proceedings, 2004, 51, 445–454.

Drivers and Pressures on Biodiversity in Analytical Frameworks

STEFAN KLOTZ

1 Introduction

Biodiversity is an important term including all aspects of the diversity of life on Earth.[1] The definitions differ, but in general it is broadly accepted that biodiversity includes genetic diversity, species diversity and the diversity of ecosystems and landscapes, including all the intrinsic ecological structures and processes like food web structure, competition hierarchies and processes like succession facilitation and inhibition.[2]

Biodiversity structures and processes are influenced and regulated by a set of extrinsic and intrinsic factors like the site condition (or classical biotic factors) and intrinsic regulating mechanisms of biodiversity itself, like population and community processes and, of course, evolutionary processes on different spatial and temporal scales.[2]

Since the Neolithic revolution and the development of agriculture, humans now are becoming the major ecological factor influencing most of the extrinsic and intrinsic factors which regulate biodiversity.[1] The process of anthropogenic biodiversity change has become more and more important and is today the major threat to biodiversity.[3]

In contrast with other environmental resources like water and air, *etc.*, biodiversity has two characters. Biodiversity is an important resource and also a working mechanism, providing different functions and services like water cleaning, erosion control, regulating matter fluxes in general, *etc.* For this reason, biodiversity loss means not only loss of species and genetic resources but also of important services or life-supporting mechanisms.[1] The study of the main influencing factors is important and must be embedded into an analytical framework explaining the relationship between humans and nature, enabling the formulation of recommendations for the management of biodiversity.[4]

Issues in Environmental Science and Technology, No. 25
Biodiversity Under Threat
Edited by RE Hester and RM Harrison
© The Royal Society of Chemistry, 2007

The state of biodiversity, biodiversity threat and loss, is well documented, not least within the IUCN reports and Red Lists on the one hand and the IPCC reports and the Millennium Ecosystem Assessment[1] on the other. The change and loss of genetic diversity, of species, ecosystems and ecosystem functioning and services must be set into an analytical framework which includes the main causes, consequences and possible responses needed to constrain or to solve the problem.[5]

In this chapter the main causes and consequences of biodiversity loss are described along with the existing frameworks needed to analyse the problem and to respond to it. Additional related and complementary material appears in other chapters of this book, in particular in Chapter 3 by Philip Hulme and Chapter 5 by Alison Hester and Rob Brooker.

2 Different Approaches to Classify the Drivers and Pressures on Biodiversity

Human impact on nature in general and on biodiversity in particular has been recognised for a long time. Different assessment methods have been developed, depending on the target parameters and on the state of the art in ecology and related sciences.[5,6] The first, more complex framework was developed through summarising the possibilities of bioindication.[6] The basic idea was to use biotic or ecosystem parameters as indicators of human impact in terms of pollution, land-use change, *etc.* Especially in the UK and Germany, specialised and complex methods have been developed to use changes within organisms, species richness, species composition and many other criteria as response variables for human pressures.[6,7] But such studies were restricted to finding correlations between the amount and duration of human impact and the reaction of biological systems. The integration of the main causes regulating the impact factors was mostly excluded. The absence of the socio-economic dimension was the main deficiency of these approaches.

2.1 Scenarios of Biodiversity Change for the Year 2100 and the Ranking of the Main Drivers

Sala *et al.*[3] modelled biodiversity change following different scenarios and ranked the importance of influencing factors like land-use change, climate change, nitrogen deposition, biotic exchange and increasing atmospheric carbon dioxide concentration. Biodiversity change was defined at the biome level following Bailey's ecoregions[8] as changes in number and relative abundance of species that occur naturally in the biomes. These anthropogenic factors are defined as drivers of biodiversity change. The changes expected by 2100 for the main biomes (arctic tundra, alpine tundra, boreal forest, grasslands, savannahs, Mediterranean ecosystems, deserts, northern temperate forests, southern temperate forests and tropical forests) are combined with the calculation of the impact of a large change in each driver on the biodiversity of each biome.[3] For the expected changes in the drivers and in the impact of changes, values from 1 to 5 were given (1 for small changes or impacts, 5 for largest changes or impacts). A large variation across biomes was found. Land-use changes were detected as the main driver of change (first rank), especially important for the tropical forests biome, Mediterranean and south temperate forests, savannahs and grasslands. Climate change was classified as of second rank and is especially important for the more northern biomes, especially the arctic, alpine and the boreal biome. Increase in nitrogen deposition was put on the third rank, mostly influencing freshwater ecosystems, northern temperate forest biome and boreal and alpine ecosystems. As the fourth main driver, global biotic exchanges were found to have the highest impacts in freshwater ecosystems and the Mediterranean biome. Rising carbon dioxide concentrations in the atmosphere will have more-or-less the same influence on all biomes. The relative biodiversity change values of the biomes give hints as to which are likely to be the most endangered biomes in future. Within the terrestrial biomes, all scenarios projected that grasslands and Mediterranean ecosystems will experience large biodiversity loss because of their sensitivity to all drivers of biodiversity change. Summarising, the main anthropogenic causes of biodiversity change are defined as drivers.

2.2   The Stress-Response Framework

Going back to the late 1970s, the Stress-Response framework was developed in parallel in Canada by Statistics Canada[9] and in Europe using the term Bio-indication.[6] Anthropogenic stresses are correlated with different responses of biological systems. The responses are grouped into different levels of biotic organisation, from biochemical reactions up to changes in the distribution of species as well as changes in communities. The Stress-Response framework was the key to indicating the impact of the main pollutants on biotic systems. One of the most widespread methods was the indication of air pollution by using lichens.[7]

Later on, this Stress-Response framework was further developed by the OECD[10,11] and the United Nations[12,13,14] and is known as the Pressure–State–Response framework (PSR).

2.3   The DPSIR Framework

Following the PSR idea, the European Environmental Agency enlarged the concept by adding and characterising drivers and impacts to form the DPSIR framework[15,16] (see Figure 1).

The driving forces are defined as human activities and processes that cause pressures. In this framework the main drivers are agriculture, forestry, industry. Generalising, the driving forces are the social and economic developments which lead to pressures on the environment. The pressures are mainly the

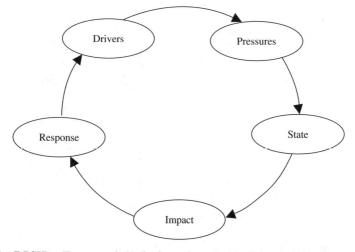

Figure 1 DPSIR – Framework (Jesinghaus, http://esl.jrc.it/envind/theory/handb_.htm (December 2006); simplified).

stresses in bioindication schemas and are defined as the direct stresses on the natural environment like pollutants to air, water and soil, radiation emissions and the intake of natural resources. These pressures influence or modify the state of biological systems like air, water and soil quality, the state of the climatic system and other observable changes in the environment. The impacts are defined as the effects of the changed environment on the systems, such as the economic sphere itself (damages in agriculture by anthropogenic environmental changes, such as an increase in floods, but also the reduction in species number in ecosystems caused by increasing nitrogen content in soils). Responses within the DPSIR framework are defined as the reaction or actions of the society to solve environmental problems like pollution prevention, setting of nature conservation schemes or more sustainable use of resources.

This framework has already been used in order to solve environmental problems in general and problems on different spatial scales from landscapes up to higher levels.[17] Svarstad *et al.*[18] dealing with discursive biases of the environmental framework of DPSIR, concluded that the problem with the framework is the lack, so far, of efforts to find a better way of dealing with the multiple attitudes and definitions of issues by different stakeholders and the general public. Many more case studies are needed to gain more experience on the use of the framework as an analytical tool and an instrument in environmental politics.

2.4 The Millennium Ecosystem Assessment Framework (MA)

The MA[1] characterises the main elements of the relationship between humans and nature differently. The main focuses are ecosystem services and human well-being, including poverty reduction, and beyond this MA targets biodiversity as a resource and a working mechanism (see Figure 2).

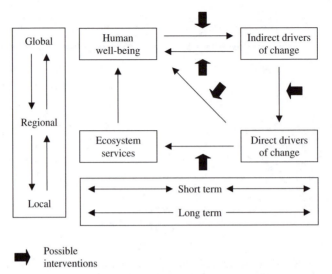

Figure 2 Conceptual framework of the Millennium Ecosystem Assessment[1] (simplified).

For that reason, the same terms used in both the DPSIR framework and the MA are defined differently in each of them. Drivers are differentiated into direct and indirect drivers. The latter are general socio-economic trends and developments like demographic changes, economic development, socio-political influences, including the forces influencing decision-making, and the quantity of public participation in decision-making. Culture and religions also can influence human behaviour and thus influence environmental change. Last, but not least, science and technology become important indirect drivers by influencing all others, especially through the use and exploitation of nature (new technologies in agriculture, fishery, forestry, and in the economic sphere in general). These indirect drivers are more-or-less the drivers within the DPSIR framework. The indirect drivers are translated into direct drivers of change called "pressures changing the state of the environment" within the DPSIR framework. The MA framework does not distinguish between the factors or processes changing environmental parameters or the state of the environment. The MA considers the most important direct drivers of ecosystem and biodiversity change to be climate variability and change, including atmospheric components like greenhouse gases and aerosol precursors. The increases in nutrient application and use, especially the three macronutrients nitrogen, phosphorus and potassium, are the main drivers for most of the ecosystems and in all biomes on Earth. Most ecosystem processes are changed by increasing the availability of nutrients.

Land conversion is the most important example of land-use changes; this is changing the land cover and reducing the size of natural ecosystems and biomes. The most important developments are the on-going deforestation within all main forest biomes and dryland degradation by overexploitation, mainly by increased grazing pressure by livestock. Agricultural conversions, the increase of arable land after deforestation and the set-aside phenomenon, are changing ecosystems. A general global problem is increasing urbanisation. The total population in urbanised areas is increasing, as is the proportion of the world's population. Urban areas are hotspots of environmental problems and the most impacted systems.

Biological invasions and the spread of diseases are seen as the fourth main direct driver of ecosystem change.[19] Biological invasions result in species loss and changes in ecosystems, which have consequences for productivity in agriculture, forestry and fishery. Human health problems could be caused by the global spread of diseases and their vectors. The economic impact of invasive species is estimated as up to 5% of world GNP.[20]

The different direct drivers act mostly in multiple ways. Interactions across all drivers are common. Land-use changes often increase the rate and extent of biological invasions. Agricultural land-use is characterised by different levels of intensity. Differing extent and degrees of intensification can be identified. The amounts of pesticide and fertiliser are, among other factors, an important indicator of agricultural land-use intensity. Additionally, land-use intensity depends on the types of crops and the crop rotation selected.[21]

Land-use conflicts between the different main land users are increasing. There are growing conflicts between land uses like agriculture, forestry, urbanisation and nature conservation. Agricultural land use and the extent of nutrient pollution are strongly related, as are urbanisation and pollution in general.

The direct drivers have a direct influence on ecosystem services. The MA defines provisioning, regulating, cultural and supporting services as important for human well-being. Within this framework, biodiversity is directly included via the ecosystem services and indirectly by the role of species and total species-richness for ecosystem functioning and services. Based on these complex inter-relations, the MA summarises the drivers important for biodiversity change simply as anthropogenic drivers comparable to the definition in the models of Sala *et al.*[3] Habitat change, loss and degradation are the main processes, followed by invasive alien species and introduced pathogens. Over-exploitation is strongly related to changes in land-use intensity and strongly combined with the first driver. All anthropogenic drivers are influenced by climate-change effects. Taking into account the spatial and temporal variation in drivers, threat processes are changing over time. It will be increasingly difficult to define the main drivers of change, as the interactions between them will become more and more complex.

3   The Main Drivers and Pressures on Biodiversity

Independent of the concept, the methods used or the framework, the main causes of biodiversity change and threat are described in a very similar way. All frameworks explain the importance of inter-relations between the drivers or pressures. Rankings depend on the target variable of biodiversity (identity of species, species number, species composition or processes important for eco-system functioning). When we define biodiversity as the target of the drivers, the drivers must be defined in a more complex way. I propose the structure in the following sections.

3.1   Land-use Changes

The most important driver, especially on the local, regional and national scale, is land-use change. Land use defines ecosystem development possible under these conditions. Land-use changes are mostly combined with loss of the previous ecosystems. Intensity of land use is strongly correlated with biodiversity, especially species richness and species composition.

It is not easy to build a hierarchy of factors within the larger frame of land-use change influencing biodiversity. However, land-use change itself can be seen as the most important and complex driver or pressure.

Land-use change is:

- the replacement of one land-use type by another type,
- the destruction of former pristine ecosystems,
- the development of new, often simpler, ecosystems,
- connected with changes in the intensity of land use, often within one land-use type.

Over-exploitation and habitat loss accompany most of the changes. Land-use change can lead to other spatial relationships between different patches of land-use types. This can be defined as changes in the landscape configuration and the degree of fragmentation, combined with problems of habitat loss and isolation of populations.

3.2   Climate Change

Climate change has two different dimensions: first, the extent of increase or decrease of quantitative climatic factors (temperature, precipitation, wind speed, *etc.*); second, the spatial and temporal variability of the factors (including the frequencies of certain events). Climate change directly influences organisms, populations, species and ecosystems and indirectly influences them by changing other drivers like land use or the immigration of alien species from warmer regions. Changes in the concentrations of greenhouse gases (methane, carbon dioxide) are the result of human economic activities which affect climate change. But climate change itself can strongly enhance the natural production of greenhouse gases, especially in former permafrost regions.

3.3   Changes in Matter Fluxes

Economic activities influence strongly the distribution of elements and chemicals on Earth, including matter fluxes. The most important cycles are the carbon, nitrogen, phosphorus, potassium, sulfur and water cycles. The human-influenced parts of the cycles are increasing permanently. I classify the increase of carbon dioxide in the atmosphere and increasing pollution by nutrients within this category. Additionally, new anthropogenic chemicals (most pesticides, pharmaceuticals) drive changes in biodiversity (*e.g.* changes in fertility[22]).

3.4 Biological Invasions

The breakdown of biogeographical barriers enables the global spread of organisms, including diseases with different evolutionary histories. The spread of alien species is an unintended large-scale experiment with significant eco-logical, health and economic consequences.[19] The increase in trade, gardening, plant breeding and the use of genetically modified organisms (GMOs) will culminate in increasing threats for biodiversity not only on islands but also on the mainlands. The genetic dimension of this process needs more detailed analyses to get a better understanding of possible future hazards. Besides the importance of invasions for the genetic level of biodiversity, the consequences on the ecosystem level are still not well analysed. On both levels the impact of alien species seems to be very important. Biological invasion is the only biological driver of biodiversity.

4 Conclusions

The different frameworks, the DPSIR framework and the Millennium Ecosystem Assessment, have both strengths and weaknesses. One problem is the different use of terms like driver, pressure, *etc.*, accompanied by possible confusion in rankings of the main factors harming biodiversity. The societal background and possible stakeholders are included differently in the frameworks. The stakeholder involvement in the DPSIR framework was discussed in detail by Svarstad *et al.*[18] These authors concluded that the DPSIR framework has some shortcomings as a tool for the establishment of communication between researchers, on the one hand, and stakeholders and policy-makers on the other. The multiple attitudes and definitions of issues by researchers, different groups of stakeholders and the general public, was identified as an additional problem.[18] In contrast with the Millennium Ecosystem Assessment, the issues of spatial and temporal scales are not included within the DPSIR framework.

The conceptual framework of the Millennium Ecosystem Assessment in-cludes both temporal and spatial scales. Ecosystem services and human well-being are the main targets of the concept. The differentiation between direct and indirect drivers and the use of the term "anthropogenic drivers" in the MA is closer to the common use of these terms in ecological sciences.[3] The indirect drivers in the MA are defined as drivers in the DPSIR framework. The direct drivers in the MA are the pressures in DPSIR.

Another problem is the different definition of single drivers. This was mainly influenced by the categories in the model of Sala *et al.*[3] using different scenarios of global change. Different factors can be summarised within the land-use category (overexploitation, fragmentation, isolation and landscape configuration). Climate change includes changes in temperatures, precipitation, wind, weather extremes and other climatic elements. The increasing carbon dioxide concentration in the atmosphere is a major cause of climate change but not the direct result. Nutrient pollution is a special case of anthropogenic pollution, with consequences for matter fluxes and anthropogenic distribution patterns of different environmental chemicals. Table 1 summarises the spread of alien plant

Table 1 Proposal for classification of drivers of biodiversity change.

Driver	Sub-driver	Consequences
Land-use change	• Overexploitation • Fragmentation • Isolation • New landscape configuration	• Habitat loss • Ecosystem loss • New anthropogenic ecosystems • Species loss • New species invasions • Changes in matter fluxes • Loss of genetic diversity
Climate changes	• Temperature • Precipitation • Wind • Weather extremes	• Changed phenology • New distribution ranges of species • New species composition in ecosystems • Genetical drift • Species loss • Species invasion
Changes in matter fluxes	• Increasing carbon dioxide concentration in the atmosphere • Pollution by nutrients • Pollution by other chemicals (pesticides and pharmaceuticals, *etc.*)	• Changes in productivity • Food webs • Competition • Toxic effects • Accumulation of toxic chemicals in organisms • Species loss • Species invasion
Biological Invasions	• Plant and animal invasions • Diseases spread • Genetically Modified Organisms (GMOs)	• Species loss • Changed ecosystem structures and functions • Hybridisation • Anthropogenic evolution

and animal species, of diseases and genetically modified organisms as sub-drivers within the biological invasions category.

Although the MA, DPSIR and Sala's[3] ranking of drivers of biodiversity and ecosystem changes produce very similar results, the aims and methods are different. To get a better ranking or categorisation better data with better spatial and temporal resolution are needed. Monitoring and assessment schemes must be better combined with or incorporated in these general frameworks.

There is an urgent need for case studies, locally and regionally, to quantify the impacts of drivers on biodiversity and human well-being. Monitoring of ecosystem functioning and services is very limited and studies focusing on the consequences of changes in ecosystem services are rare.[23]

Acknowledgements

This paper is one result of the discussions within the Network of Excellence ALTER-Net (EU-FP 6, grant GOCE-CT-2003-505298-Alternet) and with colleagues of the UFZ-Centre of Environmental Research. I thank Professor Ron Hester for help in improving the manuscript.

References

1. Millennium Ecosystem Assessment, Vol. 1–4, Island Press, Washington, 2005.
2. K. J. Gaston and J. I. Spicer, "Biodiversity: An Introduction", Blackwell Science, Oxford, 1998.
3. O. E. Sala, F. S. Chapin III, J. J. Armesto, E. Berlow, J. Bloomfield, R. Dirzo, E. Huber-Sanwald, L. F. Huenneke, R. B. Jackson, A. Kinzig, R. Leemans, D. M. Lodge, H. A. Mooney, M. Oesterheld, N. L. Poff, M. T. Sykes, B. H. Walker, M. Walker and D. H. Wall, *Science*, 2000, **287**, 1770.
4. S. Stoll-Kleemann, S. Bender, A. Berghoefer, M. Bertzky, N. Fritz-Vietta, R. Schliep and B. Thierfelder, "Perspectives on Biodiversity Governance and Management", Berlin, 2006, No. 01.
5. D. U. Hooper, F. S. Chapin III, J. J. Ewel, A. Hector, P. Inchausti, S. Lavorel, J. H. Lawton, D. M. Lodge, M. Loreau, S. Naeem, B. Schmid, H. Setälä, A. J. Symstad, J. Vandermeer and A. A. Wardle, *Ecol. Monogr.*, 2005, **75**, 3.
6. R. Schubert, "Bioindikation in terrestrischen Ökosystemen", Gustav-Fischer-Verlag, Jena, 1991.
7. M. R. D. Seaward, *Symbiosis*, 2004, **37**, 293.
8. R. G. Bailey, "Ecoregions: The Ecosystem Geography of the Oceans and Continents", Springer Verlag, New York, 1998.
9. D. Rapport and A. Friend, "Statistics Canada Catalogue", Minister of Supply and Services Canada, Ottawa, 1979.

10. OECD, "Environmental Indicators, A Preliminary Set", OECD, Paris, 1991.
11. OECD, "OECD Core Set of Indicators for Environmental Performance Reviews", OECD, Paris, 1993.
12. United Nations, "Indicators of Sustainable Development", Division for Sustainable Development, New York, 1996.
13. United Nations, "Work Programme on Indicators of Sustainable Development of the Commission on Sustainable Development", Division for Sustainable Development, Department of Economic and Social Affairs, New York, 1999.
14. United Nations, "Indicators for Sustainable Development: Framework and Methodologies", Division for Sustainable Development, Department of Economic and Social Affairs, New York, 2001.
15. EEA, "Europe's Environment: the Dobris Assessment", European Environment Agency, Copenhagen, 1995.
16. J. Holten-Andersen, H. Paalby, N. Christensen, W. Wier and F. M. Andersen, "Recommendations on Strategies for Integrated Assessment of Broad Environmental Problems", National Environmental Research Institute (NERI), Roskilde, 1995.
17. A. P. Karageorgis, V. Kapsimalis, A. Kontogianni, M. Skourtos, K.R. Turner and W. Salomons, *Environ. Manag.*, 2006, **38**, 304.
18. H. Svarstad, L. K. Petersen, D. Rothman, H. Siepel and F. Wätzold, *Land Use Pol.*, 2007, in press.
19. W. Nentwig, "Biological Invasions", Springer Verlag, Berlin, Heidelberg, New York, 2007.
20. D. Pimentel, in "Biological Invasions", W. Nentwig (ed.), Springer Verlag, Berlin, Heidelberg, New York, 2007.
21. F. Herzog, B. Steiner, D. Bailley, J. Baudry, R. Billeter, R. Bukacek, G. De Blust, R. De Cock, J. Dirksen, C. F. Dormann, R. De Filippi, E. Frossard, J. Liira, T. Schmidt, R. Stockli, C. Thenail, W. van Wingerden and R. Bugter, *Eur. J. Agron.*, 2006, **24**, 165.
22. K. Fent, C. Escher and D. Caminada, *Reprod. Toxicol.*, 2006, **22**, 175.
23. A. Balmford and W. Bond, *Ecol. Lett.*, 2005, **8**, 1218.

Subject Index

3